Technology
in the
Western Political Tradition

Symposium on Science, Reason,
and Modern Democracy

MICHIGAN STATE UNIVERSITY

Technology in the Western Political Tradition

EDITED BY

Arthur M. Melzer, Jerry Weinberger, and M. Richard Zinman

Cornell University Press

ITHACA AND LONDON

First published 1993 by Cornell University Press.

International Standard Book Number 0-8014-2724-X (cloth)
International Standard Book Number 0-8014-8006-X (paper)
Library of Congress Catalog Card Number 93-1482

PRINTED IN THE UNITED STATES OF AMERICA

*Librarians: Library of Congress cataloging information appears
on the last page of the book.*

♾ The paper in this book meets the minimum requirements of the
American National Standard for Information Sciences—Permanence
of Paper for Printed Library Materials, ANSI Z39.48-1984.

Contents

Preface

Today it is obvious to virtually everyone, as it was not as recently as thirty years ago, that there is a "problem of technology." Arguably no other comparably large theoretical issue impinges so directly on daily life—figures so prominently in newspapers, court cases, and talk shows—as does this problem in its various aspects. And now that the Cold War has ended, it is poised to become our central concern. Indeed, if the most pervasive and profound characteristic of the modern age is its ever-expanding conquest of nature, then it must necessarily follow: the problem of technology is the problem of our time.

It does not follow, however, that there yet exists—especially within the United States or within the universe of liberal democratic thought—an adequate confrontation or reckoning with this issue in its full theoretical dimensions. For the most part, this problem comes to light only as the problems of specific technologies. Is it moral and desirable to use biological engineering to alter natural species (including our own)? Who has the right to decide if a fetus will be aborted or a patient disconnected from life-support machines? How are we to deal with the potential dangers of environmental degradation or of nuclear war? Different technologies bring different problems, but beneath this apparent variety three practical questions seem to be raised by every advance in human power: Is it moral and desirable to do what this new capacity enables us to do? On whom devolves the right to decide this question? And how does one control the unanticipated or unintended consequences that may emerge from the exercise of this power? These are the problems arising from the development of particular new technologies—difficulties that we usually treat as technical problems or as policy questions to be resolved within the context of established moral and legal practices.

Not far beneath the surface of these practical issues, however, lurk much larger theoretical questions concerning the rational coherence, the psychological impact, and the moral propriety of the whole technological-scientific project. Can modern scientific rationalism really give an adequate account of the world and especially of the human world, including the human capacity for science? Is the vision of nature and of man that it imposes on us compatible with a truly human life? And does the technological project for the mastery of nature not violate some sacred or, at any rate, salutary limit on human power established by nature or God? These larger questions—which always color the concrete policy debates while escaping their direct attention—constitute the problem of technology in the deeper and more interesting sense that guides the inquiry in this volume.

To be sure, this subject is hardly new, but in our century the discussion of it has been dominated by powerful Continental critics of technology—primarily Heidegger on the right and the Frankfurt School on the left—who also stand outside and against liberal democracy, which they see as merely the political manifestation of technological rationalism. And of course there is some warrant for this view, since liberal politics and technological science began to emerge at about the same time and even, to some extent, in the same minds. They sprouted from similar premises concerning the autonomy of practical and theoretical reason from scriptural or teleological absolutes, and they were nourished by shared hopes for humanity's progressive self-improvement through its conquest of the political and natural worlds. They have also supported each other symbiotically: liberalism protecting free and public scientific inquiry, while the latter generated the technological and economic growth on which the former has depended.

Because of this intimate connection, it does seem peculiarly difficult for thinkers working within the liberal democratic tradition to confront the problem of technology in its most radical forms, as the relative silence on this topic by liberal theorists—John Rawls, Ronald Dworkin, and Robert Nozick, for example—would seem to indicate. But for the same reason it is above all necessary for liberals to address it. The problem of technology is, to a very great extent, the problem of liberal democracy.

The specific purpose of the present volume, then, is to examine the character and crisis of modern technology, and to consider what theoretical and practical resources liberal democracy might possess for dealing with it. As a glance at the table of contents will show, the approach we have adopted is primarily historical, with essays on various high points in the development of, or philosophical reflection about, technology. If the phenomenon to be examined were merely something limited and well contained, standing discreetly before us, then a more direct, analytical approach might have been appropriate. But modern technology with its attendant world view is something so pervasive and all-encompassing that we ourselves stand wholly immersed in it; and the very concepts we would employ to understand it are its own products. Therefore, our first task is just to try to separate ourselves from our subject matter—a process of de-immersion best accomplished through historical studies.

At the same time, this approach is also essential for evaluating the prevailing accounts of technology, all of which are based on sweeping historical claims. It is necessary to investigate whether, as Heidegger asserts, the Western tradition is a unified whole dominated by the technological rationalism stemming from Platonic metaphysics. Or is Western technology rather a Christian phenomenon, deriving from the biblical injunction to subdue the earth, as others maintain? Or did the technological turn occur in the Renaissance as a fundamental break with both classical and Christian thought?

More generally, we need to determine what is historically singular and what universal in the phenomenon of modern technology. Did it arise as a result of a conscious rational decision or a historical accident? Is it now susceptible to human control or is it an independent force that would rule its putative master? If the latter, what is its inner logic and destiny? If the former, is our present form of government equal to the task, or will the encounter with technology eventually compel us to abandon our commitment to liberal democracy?

Finally, the belief in a "problem of technology" itself has a history that it is necessary to explore. What is unique and what universal in our modern fears and doubts? What have we to learn from earlier resolutions of this problem—and earlier formulations of it? To what extent is our reaction against technology still colored by a technological frame of mind? In the various chapters of this volume, these questions are not always directly addressed, let alone answered, but they define the space, so to speak, within which these essays are situated.

In the Introduction to the volume, Leon R. Kass sets the general tone by specifying what we mean by the "problem of technology" and by thinking about its possible implications for liberal democracy. For Kass, we are rightly worried about technology because it provokes anew the age-old question of human happiness at the same time that it tends to undercut the validity of such questions and endangers our capacity to lead good and happy lives. For Kass, liberal democrats must rediscover the richer, nontechnological conception of liberty and dignity, original to our polity, if they are to keep their moral bearings in the modern age.

James H. Nichols, Jr. opens Part I with a study of the meaning of technology for Sophocles, Thucydides, Aristotle, and Lucretius. Nichols concludes that moderation, the virtue celebrated by almost all the ancient Greeks and Romans, may be the one lesson we need today, not just in managing the powers we discover but also regarding the large ethical responsibilities these powers tempt us to assume: responsibilities for the human race, for the globe, for the whole of nature.

Next, Stanley L. Jaki warns us against thinking that the technological project must have been a humanistic plot against religion. He argues that modern science and technology—indeed the principle of inertial motion, the very root of modern physics—can be traced to medieval sources whose Christianity was essential for discovering what the pagan Aristotle could not see. Stanley Rosen then counters with a study of the Platonic (that is, strictly philosophic) origins of the modern project. Rosen contends that Plato's poetic turn of mind—his tendency to blur the difference between contemplation and constructive making—is at least as fruitful for illuminating our technological time as is the early modern rise of the mathematical imagination. Moreover, Plato's rejection of Aristotle's belief in moral entelechy stands as a kind of paradigm for our present disagreements about the goodness of the technological project. This ancient debate concerned whether, as Aristotle maintained, our everyday moral intuitions point to what is good and complete by nature, and so to happiness and satisfaction, or whether, as Plato said, such knowledge is always elusive, escaping our grasp as it draws the soul to the restless and unquenchable desire for truth. For Plato, the philosopher longs to gaze motionlessly on the truth but is condemned to the busy, constructive activity of defending his own interest and way of life. If we add to this view the modern step of rejecting even the elusive possibility of comprehensive truth, then Plato, not Aristotle, is the figure in whose image the modern project was formed. And so when we applaud or doubt the technological project, implies Rosen, we take our stand on one side or another of the ancient divide: for happiness and natural contentment, or for the grim, constructive accommodation to desires that can never be completely satisfied.

Wilson Carey McWilliams then investigates how technological rationalism suffuses our American political life and forms of liberty. For McWilliams, America's principles of moral and political virtue spring from two contradictory sources: from earlier conceptions of nature and nature's God, and from the Constitutional Founders' more technological notions of politics, education, and moral life. McWilliams fears that in our times a balance between these different sources cannot long endure. If our liberties are to be saved from the march of technology, we must admit (with Tocqueville) that they are not rooted in the mastery and possession of things, or even in associations of our own making, but rather in divine and natural limits, whose truths must rule over our vain desires for control.

Finally, Paul A. Cantor shows us how very hard it is to stand outside of the technological frame of mind. In his study of English Romanticism, Cantor uncovers the essentially technological understanding of nature assumed by the Romantics, who so often railed against the evils of technological progress. In a sense, Cantor's Romantics danced to the tune of Rosen's Plato: both saw the worlds in which we live to be the product of our willful and poetic (and so ultimately technological) construction. The Romantics thus hold up a mirror to our own dissatisfied souls: when we lament the disenchantment of nature and sacred earth, we often speak as if they were ours to redeem in any way we see fit.

In much the same vein the essays of Part II—those on our modern technological destiny—look behind as well as ahead. G. A. Cohen shows how Marx's faith in technological superabundance kept him from addressing important questions about justice. But now that we face the dangers of technology and the limits of the earth's resources, Marxists have to face up to this old, traditional, and "superstructural" question. For Cohen, there is a Marxist account of justice, both as distributive principle and moral virtue, which can preserve individual autonomy without at the same time accepting the bourgeois principle of self-ownership. Sheldon Wolin then investigates the neo-Marxist Frankfurt School's criticism of technology and instrumental reason and argues that the fate of the Frankfurters—a kind of exile from the world of politics now dominated by technology—is reminiscent of Socrates and the tension between philosophy and political life. For Wolin, technology now has philosophy as well as politics in its deforming grasp, so that today we cannot appeal simply to philosophy to redeem political life. Indeed, we must first rescue genuine democracy if there is to be any hope for decent philosophy.

In marked contrast, Reiner Schürmann discusses Heidegger's notion of technology as the danger that "saves." It does so by way of technological nihilism, which discloses the complete uniqueness of the technological epoch and thus the utter singularity of Being, which always strives against and ruptures every attempt of technological reason to construct universal norms and standards. Technological nihilism saves by readying us to embrace the tragic contingency of every foundation and construction (such as a conception of justice, an edifice of instrumental reason, or community of philosophy or political deliberation), and by allowing us to see that this tragedy is never ours, but belongs rather to the singular happening of Being. Gianni Vattimo then makes the case for a postmodern understanding of technology, also taking his cue from the later Heidegger. For Vattimo, modern technology can have its effect both as a frame of thinking, as Heidegger understood it, and as a concrete, specific technical form. In the first sense, the advent of technological nihilism undoes the domination of metaphysical rationalism, and in the second, the era of self-conscious and multidimensional information technology replaces the linear, blind, and hierarchical order imposed by the mechanical device. In both instances the possibility of a "weak ontology" is enhanced, and, with that, the range of personal and political liberty can be expanded.

From a completely different point of view, William A. Galston responds that Heidegger's understanding of technology obscures the differences among the political orders that have faced the challenge of technology. Even though modern science casts doubt on our ability to make judgments about human values, the real and patent differences among the modern political alternatives make it possible to speak clearly about those moral principles that can best stand up to the dangers of modern technology. For Galston, these are the ends

and virtues of liberal democracy, which can be assessed by taking
seriously the everyday views of real human beings. To discern a twist
on Rosen's argument, Galston implies that we can see how technology
is good only if we side with Aristotle against the conceits of the
philosophic elite.

Jerry Weinberger then argues that technology can indeed best be
seen as a frame of thinking, and that its danger is a utopianism that
obscures the enduring forms, cleavages, and limits of political life.
For Weinberger, Heidegger's politically abstract view of technology
actually reflects one of technology's important effects. It is thus not
surprising that liberals, who put so much faith in the ameliorative
power of scientific progress, have been more influenced by Heidegger's views than might be expected. Ironically, Heidegger's influence
has come to affect recent progressive and democratic interpretations
of liberalism, which rely on the technological hope that material
abundance and the "overcoming of metaphysics" will obviate the age-
old disputes about justice. For Weinberger, such hopes are as tempting to liberals as they are debilitating to their capacity for political
prudence, and it will be important to see through them if we are to
assess what is good and bad about technology and preserve what is
best in liberal politics.

In a final essay, Arthur Melzer examines how modern technology
is unique and modernity uniquely technological. The key to an answer, he argues, lies in two fundamental features of modern culture:
the belief in progress and the disbelief in happiness, or, in other
terms, the futurism and anti-eudaemonism of modern life. An analysis of these characteristics also seems to reveal unsuspected resources
within liberal democracy for the control of technology. He concludes
by suggesting that the greatest danger of this pervasive modern force
may in fact be our growing tendency to demonize it and to reformu-
late all the issues of philosophy and politics in terms of the "problem
of technology."

The authors in this volume all agree that modern technology really
is something new, and none proposes that we can or should wish
away our extraordinary powers. They agree that technology is Janus-
faced, and even then in a double sense. As a growing system of tools
and means it is both friend and foe—within our power as helpful
genie and beyond our control as unexpected and dangerous consequence. And as an intellectual force it has on the one hand an uncanny power to surprise and upset our given worlds, and on the

Technology
in the
Western Political Tradition

be exaggerated. We have in recent decades lost our innocence about technology, and, hence, we need all the more to try to understand it. The present effort will be justified if I can gain some clarity about the question and help the reader to do the same.

Probably the most common view of our topic is something like the following: technology is the sum total of human tools and methods, devised by human beings to control our environment for our own benefit. Because it is essentially *instrumental,* technology is itself morally neutral, usable for both good and ill. There are, of course, dangers of abuse and misuse of technology, but these appear to be problems not of technology but of its human users, to be addressed by morality in general. And, besides abuse and misuse, there is a genuine problem of technology itself: the unintended and undesired consequences attending its *proper* use. Thus the problems of technology can be dealt with, on one side, by technology assessment and careful regulation (to handle side effects and misuse), and, on the other side, by good will, compassion, and the love of humanity (to prevent abuse). This combination will enable us to solve the problems technology creates without sacrificing its delightful fruits.

This view, I contend, is much too simple. It holds too narrow an understanding of the nature of technology, too shallow a view of the difficulties it produces, and too optimistic a view of our ability to deal with them—not least because this vision is itself infected with the problem of technology. That at least is what I will now try to show.

What Is Technology?

Our first problem regarding technology is to attempt to understand what technology is—no easy matter. The term itself is singularly unhelpful. According to the Oxford English Dictionary, its original English meaning, dating back to the early seventeenth century, was "a discourse or treatise [i.e., a logos] on art or the arts," or, again, "the scientific study [a logos] of the practical or industrial arts." A second meaning identifies technology as "technical nomenclature," that is, the terminology or speech—logos—of a particular art. Only in the second half of the nineteenth century was the meaning transferred to refer to the practical arts themselves, taken collectively: "His technology consists of weaving, cutting canoes, making rude weapons" (1864).

Etymologically, the term "technology" has Greek roots: *technē*, art, meaning especially the useful crafts rather than the fine arts, that is, carpentry and shoemaking rather than poetry and dance; and *logos*, articulate speech, or discursive reason. But the Greeks did not know the compound, *technologos*. As far as I can tell, the closest they came to any such notion would have had the emphasis reversed: not an account about art (a *logos* of *technē*) but an art of speaking. Rhetoric, the art of persuasive speech, was indeed a *technē* of *logos*, and, in the view of the sophists, a means of rationalizing political life free of the need for force.[1] (One could, I think, get very far in understanding the difference between the ancient polis and the modern nation-state by beginning with and thinking through the difference between technology understood as *rhetoric* and technology understood as *rationalized art and industry*, but that must be left for another occasion.)

Still, art and speech are intimately related. Both are manifestations of human rationality, of the fact that man is the animal having logos, the rational or reasoning animal. Human craft, unlike animal making, is not spontaneous or instinctive. It involves deliberating, calculating, ordering, thinking, planning—all manifestations of logos. The connection was observed succinctly by Aristotle: *technē*, he says, is a disposition or habit of *making* (as contrasted with doing), involving true reasoning *(logos)*.[2] All artful making has a manual element, to be sure, but to be truly *technical* it must be guided by mind, know-how, expertise. It is this mental and rational element that makes the various arts eminently teachable—through various "how-to-do-it" guides and manuals. (About the *disposition* to make—about why human beings *want* to make—we will speak shortly.) Following up these clues, one might think that technology is the sum of the products of craft and industry, and, even more, the sum of the know-how, skills, and other devices for their production and use.

But this is, at best, a partial view. Technology, especially modern technology, occupies itself not only with the bringing-into-being of machines and tools and other artifacts. It is centrally involved in the harnessing of power and energy—thermal, hydroelectric, chemical, solar, atomic. The drilling for oil, the damming of rivers, the splitting of atoms provide not objects of art but an undifferentiated ready resource for all sorts of human activities, in both war and peace. Indeed, according to Heidegger, this aspect of modern technology is

1. See, for example, Plato's *Gorgias* 450C.
2. Aristotle, *Nicomachean Ethics*, 1140a20.

essential and decisive. Modern technology is less a bringing-forth of objects than a setting-upon, a challenging forth, a demanding made of nature: that nature's concealed materials and energies be released and ordered as standing-reserve, available and transformable for any multitude of purposes.[3] Not the loom or the plow but the oil-storage tank or the steel mill or the dynamo is the emblem of modern technology.

Though a welcome addition, this, too, does not go far enough. For technology today is not limited to external and physical nature: technology now works directly on the technologist, on man himself. There is burgeoning biomedical technology, to this point largely harnessed by the art of healing, but in the future usable also for genetic engineering and the like. There is psychological technology, from various techniques of psychotherapy to psychopharmacology. There are abundant techniques of education, communication, and entertainment; techniques of social organization and engineering (for example, the army and the police); techniques of management (the factory or the boardroom); techniques of inspection and regulation; techniques of selling and buying, learning and rearing, dating and mating, birthing and dying, and even, God help us, for grieving. In modern times, as Jacques Ellul has persuasively argued, the technical is ubiquitous, much wider and deeper than the mechanical or the energetic. For him (as for Heidegger), technology is an entire way of being in the world, a social phenomenon more than a merely material one, characterized by the effort, through rational analysis, methodical artfulness, and correlative organization, to order all aspects of our world toward efficiency, ease, and control—to achieve the fullest control at the highest efficiency at the least possible cost and trouble.[4] Technology comprises organization and scheduling no less than machinery and fuel, concepts and methods no less than physical processes. In short, it is a way of thinking and believing and feeling, a way of standing in and toward the world. Technology, in its full meaning, is the disposition rationally to order and predict and control everything feasible, in order to master fortune and spontaneity, violence and wildness, and to leave nothing to chance, all in the service

3. Martin Heidegger, "The Question concerning Technology," in *The Question concerning Technology and Other Essays,* trans. William Lovitt (New York: Harper & Row, 1977), esp. pp. 14–17.

4. Jacques Ellul, *The Technological Society,* trans. John Wilkinson (New York: Vintage Books, 1964), p. 21.

of human benefit. It is technology, thus understood as the disposition to rational mastery, whose problem we hope to discover.

Whence comes such a disposition to mastery? What is the source of the technological attitude? Again, a terribly difficult question, hard to unravel. According to some, its deepest roots are somehow tied to human weakness: necessity is the mother of invention. Need lies behind the fish hook and the plow, fear of beasts and men behind the club and the barricade, and fear of death behind medicine. It is, according to Hobbes, the fear of violent death that awakens human reason and the quest for mastery. Of course, too much fear can enervate. According to Aeschylus's Prometheus, only when men ceased seeing doom before their eyes were they able, with his aid, to rise up from abject nothingness, poverty, terror.[5] On this view, the world's inhospitality—not to say hostility—toward human need and aspiration inspires the disposition to self-help through technology.

On other accounts, the root is not primarily human weakness but human strength: pride rather than needy fear erects the technological attitude. According to Genesis, for example, the first tool was the needle, and the first artifact the fig leaf, as shame (which is here nothing but wounded pride) moved the primordial human beings to cover their nakedness, right from the moment of their rise to painful self-consciousness.[6] Pride lies behind the technological project of the city and tower of Babel, the human race being moved by the desire to make itself a name through artful self-assertion.[7] At the beginning of the modern era, Francis Bacon, himself moved by honor and glory, calls mankind to the conquest of nature for the relief of man's estate, which project he regards as the highest and most magnificent human possibility.[8] Ambition—the desire for wealth, power, and honor—prompts many a man of science and industry; finally, the master does not seek mastery just to escape from the cold.

There are, of course, other possible roots than fear, need, and pride: for example, laziness, beneath the desire for an easier way to mow

5. Aeschylus, *Prometheus Bound*, trans. David Grene, in *Aeschylus II, The Complete Greek Tragedies*, ed. David Grene and Richmond Lattimore (Chicago: University of Chicago Press, 1956), lines 250; 437ff.

6. Genesis 3:7.

7. Genesis 11:1–9. See my essay, "What's Wrong with Babel?" *American Scholar* 58 (Winter 1988–89), 41–60.

8. Francis Bacon, *The Advancement of Learning Book I* and *The Interpretation of Nature (Proem)*, in *Selected Writings of Francis Bacon*, ed. Hugh G. Dick (New York: Random House, 1955), pp. 193 and 150–54.

the lawn; boredom, beneath the desire for new amusements; greed, beneath the desire simply for more and more; vanity and lust, beneath the desire for new adornments and allurements; envy and hatred, beneath the desire to afflict those who make us feel low. And there is also the hard-to-describe desire to do something, to make something, to order something *just to see it done*—call it curiosity, call it willfulness, call it daring, call it perversity, call it will-to-power, I think we all know the motive, even from the inside.

This analysis of the origins of the technological disposition is thus far only psychological, finding its roots deep in basic features of the human psyche. Yet this cannot be the whole story. For one thing, not all human societies would be rightly described as technological, even though all of them practice at least some of the arts. The people of ancient Israel or the early Native Americans were not technological nations, nor is technology today the ruling outlook in Iran or in much of modern India and Africa. In place of the disposition to rational mastery, these societies, and many others like them, are ruled by the spirit of reverence, or national pride, or the passion for righteousness or holiness or nobility, or even just the intense devotion to one's own traditions. And even in the rationalist West, technological society seems to have appeared only in the last two centuries, and, at its now runaway pace, only in the last seventy or eighty years. It certainly seems as if *modern* technology differs from ancient *technē*, not only in scale but even, decisively, in its nature.

Whether or not this is so can be argued at length. But one thing is indisputable. Modern technology would not be the ubiquitous phenomenon it is—or, I submit, the problem it is—were it not for modern science, that daring and stupendous edifice of still burgeoning knowledge, built only over the last 350 years on foundations laid by Galileo, Bacon, and Descartes. Says Robert Smith Woodbury in his article "History of Technology" for the *Encyclopaedia Britannica*: "For many thousands of years . . . [man's] progress in technology was made by trial and error, by empirical advance. . . . It was only toward the end of the 18th century that technology began to become *applied science,* with results in the 19th and 20th centuries that have had enormous influence."[9] A discussion of the nature of technology would be incomplete without at least a few words about modern science.

Though it is fashionable to distinguish *applied* from *pure* science—

9. Robert Smith Woodbury, "History of Technology," *The Encyclopaedia Britannica*, 14th ed. (1973), 21:750. Emphasis added.

and it makes some sense to do so—it is important to grasp the essentially practical, social, and technical character of modern science as such. Ancient science had sought knowledge of *what* things *are*, to be contemplated as an end in itself satisfying to the knower. In contrast, modern science seeks knowledge of *how* things *work*, to be used as a means for the relief and comfort of *all humanity*, knowers and non-knowers alike. Though the benefits were at first slow in coming, this practical intention has been at the heart of modern science right from the start. Here, for example, is the celebrated announcement by Descartes of the good news of knowledge "very useful in life":

> So soon as I had acquired some general notions concerning Physics
> . . . they caused me to see that it is possible to attain knowledge
> which is very useful in life, and that, instead of that speculative
> philosophy which is found in the Schools, we may find a *practical
> philosophy* by means of which, knowing the *force* and the *action*
> [N.B., not the *being* or the *nature*] of fire, water, air, the stars, heaven,
> and all the other bodies that environ us, as distinctly as we know
> the different crafts of our artisans, we can in the same way *employ
> them* in all those uses *to which they are adapted,* and *thus render our-
> selves as the masters and possessors of nature.*[10] (Emphasis added)

(Descartes goes on to enunciate the humanitarian purposes of science-based mastery: the conquest of external necessity, the promotion of bodily health and longevity, the provision of psychic peace or a new kind of practical wisdom.)

But modern science is practical and artful not only in its end. Its very notions and ways manifest a conception of the interrelation of knowledge and power. Nature herself is conceived energetically and mechanistically, and explanation of change is given in terms of (at most) efficient or moving causes; in modern science, to be *responsible* means to *produce* an *effect*. Knowledge itself is obtained productively: hidden truths are gained by *acting* on nature, through experiment, twisting her arm to make her cough up her secrets. The so-called empirical science of nature is, as actually experienced, the highly contrived encounter with apparatus, measuring devices, pointer readings, and numbers; nature in its ordinary course is virtually never directly encountered. Inquiry is made "methodical" through

---

10. René Descartes, *Discourse on the Method*, in *The Philosophical Works of Descartes, Volume I*, ed. Elizabeth S. Haldane and G. R. T. Ross (Cambridge: Cambridge University Press, 1981), p. 119.

the imposition of order and schemes of measurement "made" by the intellect. Knowledge, embodied in *laws* rather than theorems, becomes "systematic" under rules of a new mathematics expressly *invented* for this purpose. This mathematics orders an "unnatural" world that has been intellectually "objectified," re-presented or projected before the knowing subject as pure homogeneous extension, ripe for the mind's grasping—just as the world itself will be grasped by the techniques that science will later provide. Why, even the modern word "concept" means "a grasping together," implying that the mind itself, in its act of knowing, functions like the intervening hand (in contrast to its ancient counterpart, "idea," "that which can be beheld," which implies that the mind functions like the receiving eye). And modern science rejects, as meaningless or useless, questions that cannot be answered by the application of method. Science becomes not the representation and demonstration of truth but an *art:* the art of *finding* the truth—or, rather, that portion of truth that lends itself to be artfully found. Finally, the truths modern science finds, even about human beings, are value neutral, in no way restraining, and indeed perfectly adapted for, technical application. In short, as Hans Jonas has put it, modern science contains manipulability at its theoretical core[11]—and this remains true even for those great scientists who are themselves motivated by the desire for truth, who have no interest in that mastery over nature to which their discoveries nonetheless contribute and for which science is largely esteemed by the rest of us and mightily supported by the modern state.

For this reason, we must think of modern science and modern technology as a single, integrated phenomenon. It is the latter's fusion with the former that both makes it so successful and, as we shall see later, makes it such a problem.

WHAT IS A PROBLEM?

If we now know what we might mean by "technology," we need also a few words about "problem" if we are finally to address our question, "The Problem of Technology." What do we mean by "a problem"? This is no semantic game, for there is a deep difference

---

11. Hans Jonas, *Philosophical Essays: From Ancient Creed to Technological Man* (Englewood Cliffs, N.J.: Prentice-Hall, 1974), p. 48.

between thinking of something as a problem rather than, say, as a question. (Heidegger's famous essay is "The *Question* Concerning Technology.") The word "problem" comes from the Greek word *problēma*, meaning literally "something thrown out before" us. A problem is any challenging obstacle, from a fence thrown up before an armed camp to a task set before someone to be done. Problems are publicly articulated tasks that challenge us to solve them, which is to say, to do away with them as problems or obstacles. When a problem is solved, it disappears as a problem; its solution is its *dis*solution. The solution is usually a construction, something we *make*, put together from elements into which the problem is broken up or, as we say, analyzed. We model the problem into a shape convenient for such analysis and construction; as we say, we *figure* it out. Further, a problem requires a solution *in its own terms*; the solution never carries one beyond the original problem as given.

The model of such problem-solving is algebra. The equations containing unknowns are arranged, showing the analyzed elements in their constructed relations. The solutions that identify the unknowns dissolve the problem and render the equation into an identity or tautology, which invites no further thought. But this mode of thought is not confined to algebra. It is the dominant mode of all modern scientific and technological thinking, in which, as we have seen, the mind makes its perplexities operational by turning them into problems that it can then methodically solve. But, further, this mode of thought is, in fact, the dominant mode of everyone's thinking much of the time. We are always trying to figure something out, to find a way over or around our obstacles, to solve our problems. We do not like to be obstructed, to be at a loss, to be resourceless. We do not like to have problems.

What, then, does it mean to treat technology as a problem or to ask about *the* problem (or even the problem*s*) of technology? Before *whom* is technology an obstacle? More importantly, with respect to *what goals* or desiderata does technology obstruct? Is it an obstacle to human happiness or to justice or to self-knowledge? Could technology, understood as the disposition and activity of mastery, turn out to be a stumbling block in the path of the master himself? Finally, if technology is a problem, or poses a problem, what would or could be its "solution"? If the difficulties caused by technology are, in fact, only problems—say, for example, the problem of unintended side effects—we are certainly well advised to be looking for solutions. But what if the difficulties attending technology are both integral to its very being and inseparable from its benefits—like the other side of a coin? This would make technology more like a tragedy that begs for understanding and endurance than like a problem that calls for a solution. To put the point starkly: to formulate the question about technology as the *problem* of technology is itself a manifestation of technological thinking—of the desire to knock down all obstacles, even if only in the mind. To ask about the *problem* of technology, in fact, *exemplifies* it. I hope this will become clearer by the end.

THE PROBLEMS OF TECHNOLOGY

My warning about the term "problem" notwithstanding, I proceed to consider the problems of technology. Needless to say, I make no pretense of comprehensiveness. Moreover, I would not even attempt a balance sheet of benefits and harms, even for a single circumscribed technique: the task is virtually impossible and the gain in fundamental understanding slight. Finally, though I shall speak only about problems—for clearly the blessings of technology are well known to us all—I am not, I repeat, an enemy of science and technology, at least in their proper measure, and I hold no brief for romanticism, irrationality, or the good old days. In the good old days, in those good old places, I, and most of you, would very likely either have died in infancy or spent an abbreviated life in arduous toil, also persecuted for our heresies.

The first thing I would observe is that the argument about technology—or at least about the goodness and sufficiency of human art and craftiness—is very old. Nearly everyone in antiquity agreed that some form and degree of artfulness is indispensable for meeting human needs and for human living together. No arts, no cities, and if no cities, no true humanity. Rational animal, technical animal, political animal—it is all one package. The arguments concerned rather the unqualified goodness of this package and, even more, the relative importance of *technē* and of *law* (or piety) in promoting the human good. Crudely put, the argument could be stated this way: Those who hold that the biggest obstacles to human happiness are material, arising from scarcity and the stinginess and violence of nature, from the indifference of the powers that be, or (within) from disease and death, look to the arts. On this view, the inventors and bringers of the arts are the true benefactors of mankind, and are revered like the gods. The supreme example is Prometheus (literally, "forethought"), bringer of fire, with its warming and transforming power, and, through fire, all the other arts. In contrast, those who hold that the biggest obstacles to human happiness are psychic and spiritual, arising from the turbulences of the human soul itself, look instead to law (or to piety or its equivalent) to tame and moderate the unruly and self-destroying passions of human beings. On this view, the lawgivers, statesmen, and prophets are the true benefactors of mankind—not Prometheus but Lycurgus, not the builders of Babel but Moses. The arts are suspect precisely because they serve comfort and safety, because they stimulate unnecessary desires, and because they pretend to self-sufficiency. In the famous allegory of the cave in Plato's *Republic*, Socrates implies that it is the Promethean gift of fire and the enchantment of the arts that hold men unwittingly enchained, warmed and comfortable, yet blind to the world beyond the city. Mistaking their crafted world for the whole, men live ignorant of their true standing in the world and their absolute dependence on powers not of their own making and beyond their control.[12] Only when the arts and men are ruled politically, and only when politics is governed by wisdom about the human soul and man's place in the larger whole, can art contribute properly to human flourishing.

The coming of the modern technological project added a new wrinkle to this dispute. What if technology, founded upon the new science, could address not only stingy nature without but also unruly nature within? What if a science-based technology could be brought to bear on the human psyche (and human society) by means of a perfected psychophysics (and scientific political science)? Might one not eventually secure human happiness by purely rational and technical means without the need for law or force or fear of God? This was certainly part of the vision of Descartes and, it seems, one of the anticipated benefits of the mastery of nature: "For the mind depends so much on the temperament and disposition of the bodily organs that, if it is possible to find a means of rendering men wiser and cleverer than they have hitherto been, I believe that is in medicine that it must be sought."[13] Medicine, that venerable and most humanitarian of arts, will, when it is properly transformed by the new science of nature and human nature, provide at long last a solution for the

12. Plato, *Republic* 7.514Aff.
13. Descartes, *Discourse on the Method*, p. 120.

human condition. (Later, after the Enlightenment, a similar promise was also tendered in the name of a scientific anthropology and sociology.)

It is to the problems of modern technology, understood as the disposition toward a *comprehensive* mastery of nature—*and human affairs*—that I now turn. My discussion has three subparts: the problem of feasibility, the problem of goals and goodness, the problem of tragic self-contradiction.

Feasibility

The first question is whether mastery of nature is feasible. Are we really *able* to exercise control, even over our material environment? Here we face first the practically important, though theoretically less interesting, problem of the unintended and undesired side effects of machine technology: air pollution, ozone depletion, soil erosion, acid rain, toxic wastes, nuclear fallout—a list of terribly serious problems, most of which can be remedied, if at all, only by more and better technology, not less, including the techniques of monitoring and regulation. Even if the prophets of imminent ecological disaster exaggerate, theirs is an urgent cause. Indeed, Hans Jonas has powerfully argued that, thanks to the profound and global effects of twentieth-century technology, the whole scale and meaning of human action have been transformed, and a new ethic centered on responsibility for the entire earth needs urgently to be inculcated. Its new categorical imperative: "Do not compromise the conditions for an indefinite continuation of humanity on earth."[14]

Some consequences, called side effects because unwanted, are, in fact, not effects to the side. Intimately connected with the intended intervention, they must be seen, wanted or not, as *central* and integral to the whole. Consider automobility: by its very nature it entails roads and bridges; the need for fuel and the dependence on oil; the rise of steel mills, auto factories, auto workers, auto dealers, gas stations, garages, body shops, traffic laws, traffic police, parking facilities, and driving instructors; the production of noise, fumes, smog, and auto graveyards; the need for auto mechanics, safety and highway inspec-

14. Hans Jonas, *The Imperative of Responsibility: In Search of an Ethics for the Technological Age* (Chicago: University of Chicago Press, 1984), p. 11.

tors, insurance agents, claim adjusters, trackers of stolen cars, parking attendants, testing and licensing personnel, and medical personnel to deal with accidents and their human victims; and, thanks also to automobility itself, urban sprawl, homogenization through destruction of regional differences, separated lives for extended families, new modes of courtship behavior, new objects of envy and vanity, and a new battleground between parents and children appear, quite predictably, on the scene. Go assess and control the so-called unanticipated consequences of letting people rapidly move themselves about! Generally speaking, it is foolish to think that human beings can exercise extensive control over the consequences of technology, not least because the users' entire life can be altered in the process. As C. S. Lewis points out, we greatly exaggerate the increase in our power over nature that technology provides. In each generation we bequeath to our descendants wonderful new devices, but by their aggregate effects we preordain or at least greatly constrain how they are able to use them.[15]

More radical analysts of technology argue that modern technique is, *in principle,* beyond human control. Jacques Ellul, in his profound study *La Technique* (in English, *The Technological Society*), argues that what looks like a free human choice between competing techniques is, in fact, *automatically* made, always in favor of greater efficiency; that technique encroaches everywhere, eliminating or transforming all previously nontechnical aspects of life; that technology augments itself irreversibly and in geometric progression; and that technology advances as if under an iron law: "Since it was possible, it was necessary."[16] This alleged automaticity of technology and its manifold effects, if true, would certainly embarrass the pretense to mastery. Winston Churchill put it starkly over forty years ago, well before most of his contemporaries even suspected that there was any problem with technology: "Science bestowed immense new powers on man and, at the same time, created conditions which were largely beyond his comprehension and still more beyond his control. While he nursed the illusion of growing mastery and exulted in his new trappings he became the sport and presently the victim of tides and currents, whirlpools and tornadoes, amid which he was far more

---

15. C. S. Lewis, *The Abolition of Man* (New York: Macmillan, 1965), pp. 69–71.
16. Ellul, *The Technological Society,* p. 99.

helpless than he had been for a long time."[17] In this connection, let me just mention in passing the special kind of helplessness experienced by millions of people in the twentieth century as a result of modern despotism, to whose utopian programs and tyrannical successes and excesses modern technology—military, psychological, organizational, and so forth—has contributed mightily. This sobering fact reminds us that what is called "man's power over nature" is, in fact, always power of some men over other men—not only under despotisms but even when it is benevolently used and used to good effect. This thought about bad and good intentions and outcomes leads us to the problem of the *ends* of technology.

Goals and Goodness

Let us set aside questions of feasibility, and the problem of unwanted and unwelcome consequences, and think now only about the desired goals. What, on closer examination, can we say about the goals of the project for the mastery of nature?

As we have already noted, mastery of nature seeks, to begin with, man's liberation from chance and natural necessity, both without and within: freedom from toil and lack, freedom from disease and depression, freedom from the risks of death. Put positively, these goals could be called comfortable sustenance, health (somatic and psychic), and longevity. Yet these wonderful things are not yet full human flourishing; providing the conditions for the pursuit of happiness, they are not yet happiness itself. Is there, in fact, any clear notion of happiness and human flourishing that informs the technological disposition?

Further, what is the *ground* of these and other goals? Where do they come from? Does liberation from natural necessity also mean liberation from natural goals? Or does man, even qua master of nature, remain within the grip of nature, as regards his goals? Are not the goals of survival and pleasure, for example, given to man by his own nature? If so, then mastery over nature will always be incomplete: what looks like mastery will, in fact, be *service*, will be subordination to the dictates of nature spontaneously working within and

17. "The Twentieth Century—Its Promise and Its Realization," speech delivered at Massachusetts Institute of Technology, March 31, 1949, in *Winston Churchill, His Complete Speeches 1897–1963*, ed. Robert Rhodes James (New York: Chelsea House, 1983), 7:344.

finally beyond human control. Mastery will be enslavement to in-
stincts, drives, lusts, impulsions, passions.[18]

On the other hand, perhaps technology can enable man to escape
completely from the grip even of his own nature, even regarding the
goals of mastery (though in another sense, of course, man cannot
simply *be* outside of nature). If so, then mastery of nature would
mean the virtually complete liberation of man *from nature's* control.
Reason would not only be a tool, reason would create. Man would
be free for full self-exertion toward goals he freely sets for himself.
Man's goals, like the means he uses to achieve them, would be his
own projects. This alone would be genuine mastery, strictly speaking.
(Something like this view of technology seems to be the understanding and dream of Marxism.)

But what, one may then ask, would make these projects *good* and
good *for us?* Why would the setting of projects by the untrammeled
human will be anything but arbitrary? If we remember that man's so-
called power over nature is, in truth, always a power exercised by
some men over other men with knowledge of nature as their instrument, can it really be liberating to exchange the rule of nature for the
rule of arbitrary human will? True, such will might happen to will
philanthropically, but then again, and much more likely, it might not.
If human history is any guide, we already know what to expect from
the arbitrary positing of human projects, especially on a political
scale. And, even on a personal scale, is it really liberating—and fulfilling—to live under one's own will if that will, too, is arbitrary, if it
takes no guidance from what would be genuinely good for oneself?

Fair enough, you will say, but why must the mastering will be
arbitrary? Why can't reason function not only as a tool, not as an
arbitrary creator, but also as a guiding eye to discover and promulgate
those standards of better and worse, right and good, justice and dignity, and true human flourishing that would direct the use of power
and make the master not an arbitrary despot but a wise ruler? This
is, of course, what all the nice people want and what they naively
hope will happen. Indeed, the extreme trust in technology tacitly
assumes the existence or the emergence of a kind of knowledge of
genuine goals that would in fact guarantee not only the freedom but
also the *goodness* of the mastery of nature.

Don't hold your breath. Such knowledge of goals and such stand-

18. Lewis, *The Abolition of Man,* pp. 77–80.

ards for judging better and worse are not easily obtained, and they certainly cannot be provided by science itself. On the scientific view of the world, there can be no knowledge, properly so-called, of these matters, no knowledge, strictly speaking, about the purpose or meaning of human life, about human flourishing, or even about ethics. Opinions about good and bad, justice and injustice, virtue and vice have no cognitive status and are not subject to rational inquiry; they are, as we are fond of saying, values, merely subjective. As scientists, we can, of course, determine more or less accurately what it is that different people *believe* to be good, but we are, as scientists, impotent to judge between them. Even political science, once the inquiry into how men *ought* to live communally, now studies only how they *do* live and the circumstances that move them to change their ways. Man's political and moral life is studied not the way it is lived, but abstractly and amorally, like a mere physical phenomenon.

The sciences are not only *methodologically* indifferent to questions of better and worse. Not surprisingly, they find their own indifference substantively reflected in the nature of things. Nature, as seen by our physicists, proceeds without purpose or direction, utterly silent on matters of better or worse, and without a hint of guidance regarding how we are to live. According to our biological science, nature is indifferent even as between health and disease: since both healthy and diseased processes obey equally and necessarily the same laws of physics and chemistry, biologists conclude that disease is just as natural as health. And concerning human longing, we are taught that everything humanly lovable is perishable, while all things truly eternal, such as matter-energy or space, are utterly unlovable. The teachings of science, however gratifying as discoveries to the mind, throw icy waters on the human spirit.

Now science, in its neutrality to matters moral and metaphysical, can claim that it leaves to these separate domains the care of the good and matters of ultimate concern. This division of labor makes sense up to a point. Why should I cease to believe courage is good or murder is bad just because science cannot corroborate these opinions? But this tolerant division of live and let live is intellectually unsatisfying and finally won't work. The teachings of science, as they diffuse through the community, do not stay quietly and innocently on the scientific side of the divide. They challenge and embarrass the notions about man, nature, and the whole that lie at the heart of our traditional self-understanding and our moral and political teachings. The sciences not only fail to provide their own standards for human conduct; their findings cause us to doubt the truth and the ground of those standards we have held and, more or less, still tacitly hold.

The challenge goes much further than the notorious case of evolution versus biblical religion. Is there *any* elevated view of human life and goodness that is proof against the belief that man is just a collection of molecules, an accident on the stage of evolution, a freakish speck of mind in the mindless universe, fundamentally no different from other living—or even nonliving—things? What chance have the ideas of freedom and dignity, under even any high-minded *humanistic* dispensation, against the teachings of strict determinism in behavior and survival as the only natural concern of life? How fares the belief in the self-evident truths of the Declaration of Independence and the existence of inalienable rights to life, liberty, and the pursuit of happiness, to whose defense the signers pledged their lives, their fortunes, and their sacred honor? Does not the scientific world view make us skeptical about the existence of *any natural* rights and therefore doubtful of the wisdom of those who risked their all to defend them? If survival and pleasure are the only possible principles that nature does not seem to reject, does not all courage and devotion to honor look like folly?

The chickens are coming home to roost. Liberal democracy, founded on a doctrine of human freedom and dignity, has as its most respected body of thought a teaching that has no room for freedom and dignity. Liberal democracy has reached a point, thanks in no small part to the success of the arts and sciences to which it is wedded, where it can no longer defend *intellectually* its founding principles. Likewise also the enlightenment project: it has brought forth a science that can initiate human life in the laboratory but is without embarrassment incompetent to say what it means either by *life* or by the distinctively *human*, and therefore whose teachings about man cannot even begin to support its own premise that enlightenment enriches life.

We are, quite frankly, adrift without a compass. We adhere more and more to the scientific view of nature and of man, which both gives us enormous power and, at the same time, denies all possibility of standards to guide its use. Lacking these standards, we cannot judge our projects good or bad. We cannot even know whether progress is *really* progress or just change—or the *reverse.*

Tragic Self-Contradiction

Against these chilling thoughts, a small humanitarian voice protests that they are irrelevant. Surely one does not need certain knowledge of the good to know the evils of human misery—of poverty, sickness, depression, and death—and to know that they are evil. Surely much human good can be, has been, and still will be accomplished by the effort to conquer them. Surely it would be base to take lying down these beatings from stingy nature, and it can only be noble to rise up on our own behalf. These charges are surely right, as far as they go. But even this limited project for mastery is, *in principle*—that is, necessarily—doomed, not by its errors but by its very success. The master simultaneously with his growing mastery— and as a result of his growing mastery—also becomes more of a slave, not only as a result of the unintended consequences of technology but from its very victories. For the victories transform the souls and lives of the victors, preventing them from tasting victory as success.

Consider, for example, the technology of meeting needs and relieving toil. We in the prosperous West have come a long way. But have we satisfaction? Have we not, with each new advance, seen our desires turn into needs, yesterday's luxuries into today's necessities? Thanks to human malleability, even need is an elastic notion, and human desires are, it seems, indefinitely inflatable. As desires swell and turn into needs, the gap between human want and satisfaction does not close, indeed gets wider.

Rousseau captured the point beautifully, describing the costs of even the first efforts at easing the harshness of life:

> [S]ince men enjoyed very great leisure, they used it to pursue many kinds of commodities unknown to their fathers; and that was the *first yoke* they imposed *upon themselves* without thinking about it, and the first source of the evils they prepared for their descendants. For, besides their continuing thus to soften body and mind, as these commodities had lost almost all their pleasantness through habit, and as they had at the same time *degenerated into true needs*, being deprived of them became much more cruel than possessing them

was sweet; and people were unhappy to lose them *without being
happy to have them.*[19] (Emphasis added)

Moreover, new needs—especially in modern times—always create
new dependencies, often on nameless and faceless others far away,
on whose productivity and good will one comes increasingly to rely
for one's own private happiness. Finally, since even poverty acquires
only a relative meaning, since technology widens the possible range
between rich and poor, since vanity and envy compound the gap and
render it intolerable, satisfaction of need and desire becomes an ever-
receding mirage. And should governments step in to try to level out
the inequalities technology has spawned, they can do so only at the
cost of economic liberty. Bureaucracy, too, is a form of technology,
and living under it is anything but liberating. Does this look anything
like mastery?

And what of technology fueled not by need but by the fear of
death? Has it proved a success? True, fear of death inflicted by wild
animals rarely troubles the minds of city dwellers, but our haven
against the animals has become a human jungle where many walk,
and dwell, in chronic fear of assault and battery, sometimes so en-
raged by their impotence to control the surrounding violence that
they contribute to it themselves. On the international scene, we may
credit the absence of fear of foreign invasion to our superior might,
but the fear of death from modern warfare and international terrorism runs high, the unavoidable result of the fact that technologies
know no political boundaries. And even in medicine, that heartland
of gentle humanitarianism, the fear of death cannot be conquered.
The greater our medical successes, the more unacceptable is failure
and the more intolerable and frightening is death. True, many causes
of death have been vanquished, but the fear of death has not abated
and may, indeed, have gotten worse. As we have saved ourselves
from the rapidly fatal illnesses, we now die slowly, painfully, and in
degradation—with cancer, AIDS, and Alzheimer's disease. In our effort to control and rationalize death and dying, we have medicalized
and institutionalized so much of the end of life as to produce what
amounts to living death for thousands of people. Moreover, for these
reasons, we now face growing pressure for the legalization of eutha-

19. Jean-Jacques Rousseau, *Discourse on the Origin and Foundations of Inequality among Men,*
in *The First and Second Discourses,* ed. Roger Masters (New York: St. Martin's Press, 1964),
p. 147.

nasia, which will complete the irony by casting the doctor, preserver of life, in the role of dispenser of death. Does this sound like mastery?

Finally, what of those technologies for the soul, only now being marshaled, to combat depression, dementia, stress, and schizophrenia? What of applied psychology and neurochemistry, behavior modification and psychopharmacology? If modern life contributes mightily to unhappiness, can we not bring technology to the rescue? Can we not make good on the Cartesian promise to make men stronger and better in mind and in heart by understanding the material basis of aggression, desire, grief, pain, and pleasure? Would this not be the noblest form of mastery, the production of artful self-command, without the need for self-sacrifice and self-restraint? On the contrary. Here man's final technical conquest of his own nature would almost certainly leave mankind utterly enfeebled. This form of mastery would be identical with utter dehumanization. Read Huxley's *Brave New World*, read C. S. Lewis's *Abolition of Man*, read Nietzsche's account of the last man, and then read the newspapers. Homogenization, mediocrity, pacification, drug-induced contentment, debasement of taste, souls without loves and longings—these are the inevitable results of making the essence of human nature the last project for technical mastery. In his moment of triumph, Promethean man will also become a contented cow. Like Midas, technological man is cursed to acquire precisely what he wished for, only to discover—painfully and too late—that what he wished for is not exactly what he wanted. Or, worse than Midas, he may be so dehumanized in the process that he is even unaware that he is no longer truly human.

Having enumerated a number of problems of technology, can we find their common ground? What shall we say is *the* problem of technology? I would suggest that *the* so-called problem of technology is that technology is not problem but tragedy, that poignantly human adventure of living in grand self-contradiction. In tragedy the failure is embedded in the hero's success, the defeats in his victories, the miseries in his glory. The technological way, deeply rooted in the human soul and spurred on by the utopian promises of modern thought, seems to be inevitable, heroic, and doomed.

To say that technology as a way of life is doomed, left to itself, does not yet mean that modern life—*our* life—*must* be tragic. Everything depends on whether the technological disposition is allowed to proceed to its self-augmenting limits, or whether it can be resisted, spiritually, morally, politically. We close with a brief look at politics.

Technology and Liberal Democracy

Technology, though all-pervasive, is not yet the whole of life. And though the entire world, East and West, marches under its banner, it has a different meaning and perhaps a different fate under different regimes. As we have a special interest in our own, we look briefly, in closing, at technology and its relation to liberal democracy.

Many Americans today—including, one should emphasize, the critics as well as the friends of technology—share, explicitly or tacitly, the rationalist's dream of human perfectibility. But the Founders of the American Republic, though influenced by optimistic Enlightenment thought, were hardly utopians. They adopted a more moderate course. On the one hand, they knew human nature well enough not to underestimate the crucial importance of good laws (backed by enforcement), education, and also religion for the preservation of decency and public spiritedness. On the other hand, they appreciated fully the promise of science. The American Republic is, to my knowledge, the first regime explicitly to embrace scientific and technical progress and to claim its importance for the public good. The United States Constitution, which is silent on education and morality, speaks up about scientific progress: "The Congress shall have power . . . To promote the Progress of Science and useful Arts, by securing for limited Times to Authors and Inventors the exclusive Right to their respective Writings and Discoveries" (Article I, Section 8). Our Founders, in this singular American innovation, encouraged technological progress by adding the fuel of interest to the fire of genius.[20]

Yet progress for progress's sake was not their goal. Rather, by science and technology they hoped to help provide for the common defense and promote the general welfare and, above all, secure the blessings of *liberty* to ourselves and our posterity.

Given its devotion to liberty and equality, it should be no surprise that liberal democracy stands in an uneasy and mixed relationship to the technological project. On the one hand, it provides the perfect

---

20. See the essay by Wilson Carey McWilliams, "Science and Freedom: America as the Technological Republic," in this volume.

soil for technological growth. Economic and personal freedom, the emancipation of inventiveness and enterprise, the restless striving for improvement, the absence of class barriers to personal ambition, and the (fortuitous) natural wealth and expanse of our particular land have given technology perhaps the most hospitable home it has ever had. Moreover, the preservation of our liberties, no less than our general welfare, has been tied on more than one occasion to American engineering, rational planning, and methodical social organization; I refer in particular to the Second World War. As our liberal democracy becomes also a mass society, with mass culture, bureaucratization, multinational corporations, and big government, our need for and infatuation with technology increases, not only in terms of equipment but also in efforts to rationalize and administer all economic and social arrangements. If we add to this our American individualism, materialism, acquisitiveness, and all the other democratic traits so well described by Tocqueville, we see clearly the danger of the intensification of all the problems of technology as we cheerfully race our engines toward (at best) the soft and dehumanized despotism of Brave New World.

At the same time, however, liberal democracy, rightly understood, also contains the seeds of the remedy for the problem of technology, or rather for remembering that the problem with technology stems from embracing its vision wholeheartedly. We liberal democrats may not be able to stem the technological tide, but we do have the wherewithal—if we remain true to ourselves as *liberal* democrats—to keep from going under water. Everything depends on rejecting the rationalist and utopian dream of perfecting human beings by re-creating them, and on remembering that richer vision of human liberty and human dignity that informs the founding of our polity.

I point to three aspects of that vision. First, liberty means political liberty. It means self-rule, not self-indulgence. It means participating in local government and community affairs, not demanding government benefits. It means not only exercising rights and making claims but being willing to defend those rights with one's life and fortune and sacred honor if necessary. The American citizen is not a happy slave. True, prospects for active citizenship today often seem greatly attenuated, thanks largely to changes of scale and governance for which technology is largely responsible: it is not clear what citizenship means or requires in the megalopolis or in the superbureaucratic state. Yet voluntary associations abound, many with civic and public-spirited intent, and, thanks to technology, liberation from arduous toil provides many more people much more leisure to get involved—in schools, neighborhood organizations, local politics, and countless philanthropic and civic-minded activities.

Second, liberty means the freedom of private life. Homes, families, friends, parents and children dwell in a realm of sentiment and spontaneous affection, rooted in the bonds of lineage and relation, and presided over by some of the deepest strengths of our nature. Private life is where we come face to face with birth and love, death and sorrow, not merely as helpless victims but as connected, responsible, and thoughtful agents. Private life also provides liberty of worship, to acknowledge the dependence of human life—and of nature itself—on powers beyond us and not at our disposal. Technology today threatens our private life, threatens by invading our houses to make us comfortably homeless. Television usurps the place of conversation; modern conveniences diminish time spent cooperatively meeting necessity; and the contraceptively liberated sadly discover that pursuit of easy gratification rarely leads to deep and enduring intimacy. Marriages and families are notoriously unstable and children suffer, often in the midst of extraordinary advantages and opportunities. But, at the same time, we increasingly know what we are missing; people are learning that marriages and friendships need their sustained attention, and there is renewed interest in religion among the young, one suspects because of their dissatisfaction with a merely technical orientation toward life. As long as families and religious bodies affirm their essentially nontechnical dispositions in the world, a remedy against dehumanization may still be found.

Finally, liberalism means liberal education. Not just education for employment or even just education for citizenship, important though these are, but education for thoughtfulness and understanding, in search of genuine wisdom. Full human dignity requires a mind uncontaminated by ideologies and prejudice, turned loose from the shadows of the cave, free not only to solve factitious problems of its own devising but free especially to think deeply about the meaning of human existence, especially in a world overshadowed by technology, free to think also about the nature, purpose, and goodness of science and technology. In liberal democracy, and especially in liberal education, lies the last best hope of mankind.

Honesty compels me to add, straightaway, that liberal education—like both active citizenship and rich private life—is not flourishing in

American society. In our colleges and universities, genuinely liberal learning is sacrificed to preoccupation with narrow specialisms, trendy intellectual fashions, ideological disputes and indoctrinations, vocational and technological training, and just plain foolishness and frivolity. Yet nearly everywhere one can find pockets of thoughtfulness and genuine inquiry—a sound program here, a superb teacher or two there—beacons of light that attract and engage serious and eager students. This volume is itself the product of one such pocket, which has summoned us, as authors and readers, to think deeply about the most important human matters. It should provide some small comfort, in the face of the dreary human prospect I have presented, to observe that we are free, able, and encouraged by our society and our universities to question and *think* about it together.

Suggestions for Further Reading

Bacon, Francis. *The Advancement of Learning, The New Organon,* and *New Atlantis.* In *The Works of Francis Bacon,* ed. J. Spedding, R. L. Ellis, and D. D. Heath. London: Longmans, 1857–1870.

Descartes, René. *Rules for the Direction of the Mind,* in *The Philosophical Works of Descartes, Volume I,* ed. Elizabeth S. Haldane and G. R. T. Ross. Cambridge: Cambridge University Press, 1981.

Kass. Leon R. *Toward a More Natural Science: Biology and Human Affairs.* New York: Free Press, 1985.

Kennington, Richard. "Descartes and Mastery of Nature." In *Organism, Medicine, and Metaphysics,* ed. S. F. Spicker. Dordrecht, Holland: D. Reidel, 1978, pp. 201–23.

Rousseau, Jean-Jacques. "Discourse on the Arts and Sciences." In *The First and Second Discourses,* ed. Roger Masters. New York: St. Martin's Press, 1964.

PART I

The Emergence of Modern Technology

1 Technology in Classical Thought

JAMES H. NICHOLS, JR.

A demonic force capable of vast good or of catastrophic evil—this seems to me a fair summary of our modern view of technology. To state this view most simply: technology produces an ever-increasing abundance of life's material goods, continual progress in combating diseases, and an unending stream of innovations for our communication, entertainment, and even instruction. But the technological forces behind such blessings are no less productive of evils: the terrible weapons of war reach unprecedented levels of horror.

This simple ambiguity about technology as productive alike of peaceful goods and warlike terrors has solid descriptive value up to a point, but reflection or experience cannot fail to complicate the picture. Increasingly terrible weapons of war do indeed seem more evil, but has not their excessively terrifying quality contributed to their not being used, to a stabler peace, along the lines of hopes expressed by thinkers like Kant?[1] On the other hand, do not the peaceful blessings of technology lend themselves to possible abuses or bring about other bad consequences? Cannot the vastly increased powers called into being by technology provide new ways for some to manipulate others, with or without frank violation of their rights?

1. Immanuel Kant, "Idea for a Universal History from a Cosmopolitan Point of View," especially the Seventh and Eighth Theses, in *On History,* ed. Lewis White Beck (Indianapolis: Bobbs-Merrill, 1963).

27

May not easy abundance and variegated entertainments divert people from serious pursuits and truly human vocations?

Today both sides of technology's ambiguous promise tend to be
stated with unusual extremism: the hopes and the fears connected
with technology are strained ever tighter. On the side of fears, of
course, the area of newest concern can be briefly evoked by the words
"ecology" and "environment." Our technological way of life is having
wholly unprecedented effects on the entire world. The exhaustion of
certain natural resources, the emission of pollutants, the altering of
global climate through the greenhouse effect—these events may be
preparing some ultimate catastrophe for the human race. The whole
world, or as a recent book would have it, nature itself, is at risk.[2]

Hopes concerning what the new forms of technology are bringing
about nonetheless soar as high as or higher than ever. Somehow technology can be believed to be the root cause underlying the definitive
victory of liberal democracy over communist totalitarianism. Montesquieu hoped that modern commerce and finance would impose such
restraints on the conduct of princes as to cure them of Machiavellian-
ism.[3] Today's even higher-flying expectation is that the information
and computer revolutions will compel all regimes, through economic
self-interest, to extend freedoms and provide free markets to their
citizens. A former chief of Citibank argues that governments cannot
control the sorts of activities that involve people working with powerful computers and databases. The telecommunications revolution defeats the claims of socialism: "Technology is on the side of freedom
and of openness of markets."[4]

When I turn from contemporary discussions of technology to the
ancients' reflections on the arts (technai), I am struck by two things.
First, I find a similar ambiguity about the arts: for the ancients as for
us, the arts hold forth both advantages and dangers. But second,
both sides of the ambiguity are understood and presented within
narrower or more moderate limits: the hopes connected with the
arts show no extravagance; the fears of danger fall far short of being
apocalyptic. If our own thinking about technology is at all dazzled or
bewildered by the prevailing extreme alternatives, of destructive fiend

2. Bill McKibben, *The End of Nature* (New York: Random House, 1989).

3. Montesquieu, *The Spirit of the Laws*, bk. 21, chap. 20, trans. and ed. Anne M. Cohler et
al. (Cambridge: Cambridge University Press, 1989), pp. 387–90.

4. Walter Wriston, "Irresistible International Revolution," *Fortune*, July 3, 1989, pp. 66–67.

or rescuing angel, perhaps reflection on calmer ancient perspectives may help us to attain some further clarity on what is at issue.

THE ARTS, DEATH, POLITICS, AND PIETY: SOPHOCLES' "ODE ON MAN"

The justly famous choral ode on man in Sophocles' *Antigone* expresses, in the brief compass of four strophes, the ambiguous value of humankind's capacity to devise various arts. The Chorus first speaks with admiration of this capacity and its achievements, then reflects on its equivocal tendency to produce noble or base things, and finally evokes laws and piety as the proper guides for human beings in their actions, artful or otherwise. But a closer look at the details and the context of this ode reveals deeper and more complex levels of meaning.

To facilitate the discussion, let me first present as literal a translation of the ode as I can.[5]

[First Strophe]

Many things are awesome, and nothing comes along more
    awesome than man.
This thing, in winter's stormy southwind, travels
across the gray sea,
passing through beneath engulfing waves.
And the highest of the gods, Earth,
imperishable and unwearying, he wears away,
with plows winding back and forth year after year,
turning up the soil with the offspring of horses.[6]

[First Antistrophe]

The light-minded tribe of birds he ensnares and leads off,
and the nations of savage beasts and the briny natural growth of
    the sea,
in the net's meshy coils,

5. I have used the text and helpful notes of Sophocles, *Antigone,* ed. Martin L. D'Ooge (Boston: Ginn, 1898). I have often been guided by Elizabeth Wyckoff's fine translation in *Sophocles I,* ed. David Grene and Richmond Lattimore (Chicago: University of Chicago Press, 1954).

6. Since horses are mentioned by name in the first antistrophe, horses' offspring here probably refers to mules (so the Scholiast, quoting Homer's *Iliad* 10.352f., cited ad loc. in D'Ooge's edition).

this all around thoughtful man.
He masters with devices the out-door
mountain-climbing beast, and the shaggy-necked
horse he leads under the neck-encircling yoke,
so too the untiring mountain bull.

[Second Strophe]

Voice and wind-quick
thought and dispositions toward laws of the town he has taught
    himself, and
to escape the shafts of inhospitable frosts under the open sky
    and of driving rains,
all-resourceful. Without resources he goes toward nothing
future: only escape from Hades will he not bring in for himself.
He has figured out for himself escapes from diseases without
    devices.[7]

[Second Antistrophe]

Having this wise thing, the devising
of arts beyond hope, he moves[8] sometimes toward the bad,
    sometimes toward the noble.
Weaving in the laws of the land[9] and the oath-sworn justice of
    the gods,
High is his city. Without a city is he who resides with
what is not fine, on account of daring. May he who does such
    things
not share my hearth nor think alike![10]

"Polla ta deina," the ode begins: "Many are the things that are
deinos [terrible, awesome, uncanny, clever]," and nothing is more so
than man. The attractive translation "Many are the wonders" obscures the moral ambiguity that is present from the beginning of this
ode. The adjective *deinos* comes from the root *deos*, meaning fear.
Something *deinos* is dreadful or terrible; a person who is *deinos* can
also be "terribly clever," as Socrates was described by his accusers as
being *deinos legein*, "terribly clever at speaking." Cleverness *(deinotes)*,
according to Aristotle,[11] when used for a noble goal, is praised; when

---

7. That is, diseases formerly without remedial devices.
8. More literally, "creeps."
9. *chthonos*, of earth.
10. *ison phronein:* to think the same, sc. politically; that is, to be of the same political
persuasion or party.
11. Aristotle, *Nicomachean Ethics* 6.12; 1144a23–29.

used for base purposes, it is condemned as knavery or criminality (*panourgia,* "doing everything," stopping at nothing).

The moral ambiguity of man's cleverness is emphasized by the context of this ode. Creon has announced a decree forbidding anyone, on penalty of death, to bury the slain traitor Polynices. A guard comes in to announce terrible things *(deina):* someone has carried out the forbidden burial rites. Creon vehemently rejects the Chorus's suggestion that the deed was god-sent, given that Polynices died fighting against his city and its gods; rather, Creon attributes the deed to his enemies in the city, believing that they have bribed the guards. From that thought he turns to a general denunciation of the effects of money, silver, on men. It destroys cities and drives men from their homes. It leads men to play the rogue or criminal (*panourgia* again: doing everything, stopping at nothing, knowing no restraints or limits) and to be conversant with unholiness in every deed *(pantos ergou).* From the standpoint of a political ruler in place, Creon narrowly[12] attributes knavery, *panourgia,* to love of gain. The Chorus takes the broader view of an ethically neutral cleverness, which produces both desirable improvements for human life and base deeds; the connection of their ode with its immediate context must be something like an unstated reflection that artful human cleverness and daring must have been involved in burying the corpse in violation of the ruler's decree (which Creon himself calls not decree but law).[13]

The fear- or awe-inspiring aspect of human cleverness arises from its connection with boldness or daring. The first example given by the Chorus presents man braving wintry storm and roaring waves to cross the sea: men leave their natural place on land, moving beyond apparent limits. Second, with their plowing, men wear away indestructible and tireless Earth, highest of the gods; the almost oxymoronic "wearing away" of "unwearying" earth surely conveys more than a hint of hubris or impiety.[14]

The ode's first antistrophe displays man's mastery over the world's other animals: his hunting of light-minded birds, wild beasts, and the sea's fishes; his taming and use of horses and bulls. The first half of the ode thus presents four areas of man's artful development,

12. Too narrowly: Antigone has already imagined herself "committing the crime *(panourgia)* of piety," v. 74.
13. At verses 192, 449, 481.
14. Joan V. O'Brien, *Guide to Sophocles' "Antigone"* (Carbondale: Southern Illinois University Press, 1978), pp. 338–39.

through which he exercises daring or control in his relationship with
sea, earth, and animals whose habitats are air, land, and sea. The
next strophe speaks of man's teaching himself speech, thought, dispositions to follow laws of the town, housing for protection against cold
and rain, and medicine against diseases. These nine developments
invite comparison with the nine arts Prometheus claims to have
taught mankind in Aeschylus's play.[15] Noteworthy for their mention
by Prometheus but absence here are astronomy, divination, and the
mining and use of metals (called "benefits to men that lay concealed
beneath the earth").[16] Here in Sophocles' ode, these arts that men in
their cleverness have *taught themselves* dare to go beyond narrow natural limits but do not rise skyward toward the domain of Zeus or delve
deep under the earth. Indeed the one unbreachable limit on man's
artful daring that the Chorus explicitly names is Hades beneath the
earth. Having mentioned man's devising of housing, the Chorus calls
him all-resourceful, with means for escaping even diseases once irre-
mediable, yet unable to flee from Hades.

At this point, to conclude the whole ode, the Chorus reformulates
the sum of man's cleverness and makes fully explicit the moral ambiguity of this trait—its being used for both good and evil. The Chorus
sums up man's cleverness, something ethically neutral, as the devising of the various arts; their good or bad use depends on their being
guided by the laws of the land and the gods' justice. Their evil use
is connected with daring, here used pejoratively only (though elsewhere it can be neutral or even praiseworthy).

Man's artful action needs to be guided by the laws of the land. But
at once a problem arises: the Chorus has just spoken of man's teaching himself the dispositions of town laws. That is, politics with laws
seems to be one of the arts that human cleverness devises, but can
the humanly devised art of politics effectively impose guidance on
the human use of cleverly devised arts? That the Chorus uses the
phrase "laws of the land" (*chthonos*, of earth) instead of laws of the
city or of the town partly obscures this problem, fostering the image
of laws that belong to the land, laws not necessarily of merely human
devising.[17] Enlarging on its sense that something beyond human

15. Seth Benardete, "A Reading of Sophocles' *Antigone*: I," *Interpretation* 4 (Spring
1975), 189.

16. Herbert Weis Smyth, trans. *Aeschylus I* (Cambridge: Harvard University Press [Loeb
Library], 1946), p. 259.

17. This line of thought inevitably calls to mind Socrates' devising of a noble lie for his
best city (Plato's *Republic*, bk. 3, end).

making is needed to guide human devisings, the Chorus adds "the oath-sworn justice of the gods" to laws of the land; humans need the guidance of something higher, beyond their own control. The mention of laws of the earth/land reminds us of the earlier mention of Earth, the highest of the gods, who though indestructible is somehow violated by man's cleverly devised art of plowing. At the very moment when the Chorus refers to what guides and restrains man, we are also reminded of how incorrigibly human cleverness seeks to transcend every limit.

Laws and the gods' justice can limit human action to what is noble and what supports the life of the city. Hades—death—beyond the reach of human daring, places an inescapable limit on all human actions and devisings. How are these two kinds of limiting related? Must they not be brought into some sort of harmony with each other if human beings are to be well directed? Now politics, as exemplified by the ruler Creon, does of course seek to control the threat of death; Creon's edict is to be enforced by the death penalty. But "Hades" evokes, beyond mere death, the fate of the corpse and soul thereafter. Humans have no arts or devices to escape Hades, but they have burial rites believed to be crucial for the soul's fate. Antigone is determined at all costs to perform such rites for her brother; she will "do every deed" (panourgein), "stop at nothing" to accomplish the pious obligation (v. 74). She is devoted, with daring if not cleverness,[18] to the obligations relating to Hades beneath the earth beyond reach of any human arts. These obligations—laws—are supported by Justice (Dike) dwelling with the gods below (v. 451). They are of more ancient standing—Antigone believes them eternal—than political laws, being addressed chiefly to members of the family, a human institution that predates the arts.

Creon, as political ruler, endeavors to extend his political artfulness beyond decrees or laws with death penalties. Since Eteocles, fighting in defense of Thebes, and Polynices, fighting against it, have died at each other's hand, do not the importance and dignity of the city and its gods require signal punishment for Polynices? But Creon, in forbidding burial for Polynices, seeks to break through the limit that keeps human art from interfering with the realm of Hades below. As Tiresias asks, "What use to kill the dead a second time?" (v. 1030). To Creon he proclaims and prophesies:

18. The Chorus suggests that she has been caught in folly, aphrosyne (v. 383).

You have confused the upper and lower worlds.
You sent a life to settle in a tomb;
you keep up here that which belongs below
the corpse unburied, robbed of its release.
Not you, nor any god that rules on high
can claim him now.
You rob the nether gods of what is theirs.
So the pursuing horrors lie in wait
to track you down. The Furies sent by Hades
and by all gods will even you with your victims.[19]

Politics, an art of human devising, is a limiting ordering of man's actions through laws. But human daring, including its artful manifestations, always seeks to go beyond any set limits. To limit human beings successfully, politics must therefore also rest on something beyond human power—the gods and piety. Politics depends on human arts, but also on a realm of things holy that sets limits to arts; the political community stands somehow between the arts and piety. But as an art, politics shows the tendency, endemic to human daring, to move beyond established limits. Political wisdom demands, above all, moderation, by which politics can limit itself, especially with respect to matters of piety that reflect an overall ordering of the cosmos by more than human power.

Thoughtful moderation in Sophocles appears necessarily to rest on human convictions about the gods, at least if moderation is to be politically effective. In Thucydides' history, we see that concerns with the divine normally permeate political practice, but we also observe how piety gives way before harsh necessities or the allure of empire.

THE ARTS, POLITICS, AND THE ART OF POLITICS IN THUCYDIDES

In Thucydides' *History of the Peloponnesian War*, the development of human arts first appears in a mainly progressive light, in the section of Book 1 traditionally called his "archaeology," that is, his reasoned account of ancient things. Its context is to support the claim stated at the very beginning of the history that the Peloponnesian War was the greatest motion (*kinesis*) of the Greeks, and hence worthy of his

---

19. Verses 1068–77; Elizabeth Wyckoff's translation.

efforts to investigate and record it. In thus showing the weakness of ancient times, Thucydides presents an account of how developed Greek civilization emerged out of an earlier barbarism. The arts play an important role in these developments, from nomadic or frequently migratory groups to settled cities with walls, longer-term agricultural investment, naval commerce, more impressive military and naval forces, and generally more comfortable lives.

Interestingly, the one art discussed in greatest detail by Thucydides is the same as that first mentioned by Sophocles' Chorus: navigation. From Minos' development of a navy that suppressed piracy throughout the Hellenic sea to the Trojan and Persian wars and Athenian development of a naval empire, Thucydides elaborates naval innovations more than any other type. And at the beginning of the war, it is the navally oriented Athenians who occupy themselves with continual innovations in marine warfare and in other matters. The Corinthians (1.71.3) recognize this as an Athenian advantage over Sparta. In urging the Spartans toward war with Athens, the Corinthians point out how old-fashioned Spartan practices are, and add: "Just as in art [*technē*], it is necessary that things that arise later [*ta epigignomena*] always prevail. It is true that when a state is at rest the established practices are best left unmoved, but when men are compelled to enter into many undertakings there is need of much improvement in art [*epitechnesis*]."[20]

But we must reflect further on the context of this progressive historical account. It certainly shows the growth of Greek power toward the greatest events, but this greatest war, Thucydides tells us from the start, caused an unprecedented degree of suffering and disasters: "Never had so many cities been taken and left desolate. . . . Never had so many human beings been exiled, or so much human blood been shed, whether in the course of the war itself or as the result of civil dissensions" (1.23.2). This context must surely call into question any simple notion of progress. Must one not rethink the value of this increased power and innovative dynamism, given its outcome?

One of Thucydides' characters, Diodotus, provides some interesting material for such rethinking. The Athenians had defeated their rebellious allies, the Mytilenaeans; in harsh anger—for Athens had allowed them to retain more independence than other allies—the Athenians decided to kill all the adult males and to enslave the

---

20. Thucydides, *History of the Peloponnesian War*, ed. and trans. Charles Forster Smith, 4 vols. (Cambridge: Harvard University Press [Loeb Library], 1962). My quotations of Thucydides' text are from Smith's translation, with occasional modifications.

women and children. The next day Diodotus was among those who
argued for milder measures, mainly on the grounds of their superior
efficacy for maintaining Athens' empire. The death penalty has been
prescribed for many offenses, he argues, yet individuals, and whole
cities, nonetheless run the risk of committing them. Mankind has
tried ever more severe penalties to try to curb evildoing. Probably,
he says:

> in ancient times the penalties prescribed for the greatest offences
> were relatively mild, but as transgressions still occurred, in course
> of time the penalty was seldom less than death. But even so there
> is still transgression. Either, then, some terror more dreadful *(dei-
> noteron)* than death must be found, or we must own that death at
> least is no prevention. Nay, men are lured into hazardous enter-
> prises by the constraint of poverty, which gives them daring *(tol-
> man)*, by the insolence and pride of affluence, which makes them
> greedy, and by the various passions (3.45).

While the arts have advanced in the present time, human daring
and insolence always threaten order, and over time the resources of
punishment are exhausted. We see from another point of view why
a political ruler such as Creon has reasons to want command over a
punishment more terrible than death. And older times, while in some
ways more primitive, look in other respects less harsh than the
present.

Diodotus's discussion of crimes and punishment is occasioned by
the Mytilenaeans' daring to revolt from Athens. But what about the
Athenians themselves? They appear in Thucydides' history as innovative in the arts and daring in their acquisition of empire. Athenian
politics is far from being the moderate ordering of the Sophoclean
paradigm that would limit human artful daring. On the contrary,
Athenian politics itself becomes an art—daring and innovative. This
aspect of the Athenian mode of conducting politics finds most powerful expression in Alcibiades' argument for the Sicilian expedition, an
argument that wins out over the case for restraint stated by the cautious, moderate, pious general Nicias.

According to Alcibiades, Athens cannot set limits to her empire;
we must, he says, hold on to some, plot against others, and beware
of falling under the rule of some if we fail to exercise rule over others
(6.18.3). Most interestingly, he likens political imperial rule to other
arts: "Consider that the city, if she remains at rest, will, like anything

else, wear herself out upon herself, and her skill (*episteme*, science or art) will grow old, whereas, if she is continually at conflict [or in competition] she will be always adding to her experience *(empeiria)"* (6.18.6). The later disastrous defeat of Athens' Sicilian expedition does not disprove the soundness of Alcibiades' argument here. But surely we must see something questionable about a constant striving for imperial expansion without limits. Alcibiades himself (6.18.7), while maintaining that a change from Athens' accustomed activities, habits, and laws would lead to ruin, nonetheless concedes that the Athenian way of politics may not be best. When politics becomes a constantly innovating art, can anything but disaster impose limits?

INNOVATION IN ARTS AND POLITICS: ARISTOTLE'S ANALYSIS

Like Thucydides, Aristotle begins his *Politics* with an initial sketch of political development that seems basically progressive. As human beings move from the limits of family life to the broader scope of the village and at last to the self-sufficiency of the *polis*, the political community, they move toward a more complete good in accordance with their nature. The scattered life of ancient times, essentially depicted by Homer's presentation of the Cyclopes, holds no special charm for Aristotle. While the development toward the *polis* is in accord with a natural human impulse, nonetheless "the one who first constituted [a *polis*] is responsible for the greatest of goods."[21] Political life had to be invented by its first practitioner.

It is surprising to observe soon after in Book 1 of the *Politics* that Aristotle presents a very different perspective. In the context of examining economics (*oikonomike*, expertise concerning the community of the household or family, whose origin antedates that of the city), he examines the various arts of acquisition considered by some as central to household management or economics.[22] Aristotle mentions five ways to acquire a livelihood that seem natural enough and regarding which he raises no moral problem: they are "the way of life of the

21. Aristotle, *Politics*, trans. Carnes Lord (Chicago: University of Chicago Press, 1984), 1.2.15; 1253a30–31.
22. At the beginning of the *Nicomachean Ethics*, Aristotle himself, presumably stating the conventional view rather than his own final position, asserts that the goal of household management is wealth.

nomad, the farmer, the pirate, the fisher, and the hunter" (1.8.8; 1256a41-b1). Subsequently emerging are exchange, the invention and use of money, relatively simple commerce, more artful commerce, and ultimately a financial art that involves lending money at interest. Concerning the most sophisticated of these arts, as contrasted, for instance, with piracy, Aristotle *does* express some moral reservations. This sort of business expertise seems to break with nature by aiming at a goal of limitless wealth, beyond the moderate amount adequate for the self-sufficient good life. True economics aims at or presupposes a limited amount of useful wealth; business and financial expertise pursue unbounded gain.

How do the progressive development toward the political community and the questionable development toward contemporary business and finance relate to each other?[23] Politics, of course, presupposes considerable development of the arts of production and exchange. But politics in Aristotle's conception aims at a finite goal, the best way of life for a self-sufficient community of human beings; as the comprehensive human association, politics rules so as to limit and guide the subordinate human activities. Natural ways of acquiring sustenance (even the violent way of pirates, raiders, or brigands) can fairly readily be seen as directed to a limited natural goal. It is, above all, artful commerce and finance that lead men into the pursuit of unlimited acquisition. Business and finance, still more than war, it seems, must be restrained by thoughtful moderation if the political community is to pursue its proper goal in the right manner.

In the context of examining a proposal by Hippodamus to confer honor on those who discover something advantageous to the city, Aristotle presents a famous statement on the difference between politics and the arts, with respect to the value of innovation in each domain.[24] In the sciences, he states that advantageous changes have been made in medicine and gymnastics, and also in the arts and capacities generally. Politics is among these, and Greek laws have progressed beyond the barbaric usages that once prevailed, such as carrying weapons and purchasing wives. Earlier people, whether survivors of some cataclysm or born from the earth, were probably no

23. Wayne Ambler, "Aristotle on Acquisition," *Canadian Journal of Political Science* 17 (September 1984), 487–502, presents an illuminating discussion of these matters.

24. My discussion draws substantially on Leo Strauss, *The City and Man* (Chicago: Rand McNally, 1964), especially pp. 21–25.

wiser than ordinary or even simple folk today. Aristotle admits all of
this, concluding that it proves that one should change some laws
sometimes.

Politics, however, which operates at its best through laws, differs
from the arts in that "the law has no strength with respect to obedience apart from habit, and this is not created except over a period of
time. Hence the easy alteration of existing laws in favor of new and
different ones weakens the power of law itself" (2.8.24; 1269a20–24).
The arts involve habit to a degree, but in them knowledge plays a far
more predominant role than it does in law-abiding action. Hence arts
can be improved by innovations in knowledge with less cost stemming from associated changes in habit. Politics must rule comprehen-
sively over the activities of men, including their practice of the various
arts, and therefore stands in need of knowledge that is superior by
virtue of its comprehensiveness. But politics to attain its end must
exercise rule over human life through laws, in which the effective
capacity of knowledge is always limited by the requirements of habituation. With this situation in mind, one expects that the constantly
innovating arts will always have some tendency to escape political
control. But politics cannot reasonably ban the arts or even forbid all
innovations, since the city's way of life depends in some obvious
material respects on considerable development of the arts. And the
most basic goal of self-defense requires the political community to
keep up with innovations in the military arts.

For Aristotle, the relation of the arts to politics is complex: partly
mutually supportive and partly antagonistic. Thoughtful moderation
and prudence are needed for this relation to be properly balanced.
The city depends on the arts, including the arts of acquisition, but
the good life for citizens requires that devotion to acquisition be limited by the virtue of moderation. The innovative tendency of the
arts must not be extirpated but rather tempered by a comprehensive
political prudence that recognizes the true character (and limitations)
of law. Whereas for Sophocles moderation appears essentially linked
to piety, Aristotle seems to envisage prudence and moderation that
emerge from thoughtful reflection on the experienced world of human activities. Human passions, together with the innovative dynamic of the arts, always tend to break through the limits that
prudence and moderation would set; the latter may receive some

reinforcement from a reasoned account of nature, within which human nature and human affairs have their place.

THE EPICUREAN PERSPECTIVE OF LUCRETIUS

The Epicurean poet Lucretius presents a more explicit and more
detailed critique of progress in the arts than anything to be found in
the other authors discussed thus far. Since he also presents some
powerful statements in praise of the arts, we find an ambivalence
similar to what we have noted in others. In the case of Lucretius, this
ambivalence can be clarified relatively easily. The arts tend to be
praised when viewed as preconditions for or steps toward the ultimate elaboration of Epicurean philosophy, through which the most
complete happiness possible for human beings can be achieved. The
arts are criticized for their effects on most people in their connection
with war and immoderation.

Let me begin to elaborate this distinction by first looking at two
passages in which Lucretius celebrates the arts. In the context of
arguing that our world is perishable, Lucretius suggests that our
natural world is of rather recent origin. Among the reasons he gives
for this belief is that "even now certain arts are being perfected, even
now they are growing; much now has been added to ships; but a
while ago musicians gave birth to tuneful harmonies. Again, this
nature of things and reasoned account are but lately discovered, and
I myself was found the very first of all who could turn it into the
speech of my fatherland" (5.332–337).[25] This passage, also notable for
probably being Lucretius's least modest, sketches a picture of overall
improvement and also leaves open the possibility of reasonable hope
for further progress.

Similarly, the end of Book 5 reviews and adds to his earlier account
of human development as follows:

> Ships and the tilling of the land, walls, laws, weapons, roads, dress,
> [and] all things of this kind, all the prizes, and the luxuries of life,
> one and all, songs and pictures, and the quaintly wrought polished
> statues, use and therewith the inventiveness [*experientia*] of the ea-

---

25. I quote the translation (to which I have made occasional alterations) given in *Titi
Lucreti Cari de rerum natura libri sex*, ed. Cyril Bailey, 3 vols. (Oxford: Oxford University
Press, 1947).

ger mind taught them little by little, as they went forward step by step. So, little by little, time brings out each thing into view, and reason raises it up into the shores of light. For they saw one thing after another grow clear in their mind, until they reached the topmost peak of the arts (5.1448–1457).

Despite his earlier detailed and harsh criticism of artful progress, for the moment Lucretius forgets its bad aspects, as he leads up to his final praise of Epicurus in the beginning of Book 6. There he celebrates Athens as the source to which we owe grain, laws, and the sweet solaces of life provided by Epicurus. From the point of view of the highest, that is, most pleasant, human possibility, the arts and political laws look beneficial: they provide the material conditions, order, leisure, and partial knowledge of various aspects of nature that must precede discovery of genuine human wisdom.

While providing an expansive account of how human affairs developed, however, Lucretius presents quite another view: the arts are productive of evils because of their connection with the horrors of war and their growing from, and in turn increasing, human immoderation. (This immoderation itself, in Lucretius's analysis, is ultimately rooted in our fear of death, our unwillingness or incapacity, prior to our serious study of true philosophy, to face up fully to the truth about our mortal condition.)

In his detailed discussion of these developments, Lucretius indicates but downplays the fact that many discoveries in the arts clearly benefit us (those that protect us from wild animals, rigors of climate, scarcity of food, and so on). His critique, on which he places the main emphasis, is that the developments of the arts do not really add to our pleasure. What is at hand pleases us most and seems excellent unless we have previously known something sweeter. Something better found later on generally tends to destroy the old things and to change our feelings toward them. Lucretius unfolds this fundamental moral critique in the immediate context of discussing the development of music, as peaceful an art as one can readily imagine, and concludes: "Therefore the human race always toils idly and in vain and consumes its lifetime in empty cares. No wonder, for it has not learned what the limit of possession is nor at all how far true pleasure can increase. And that [ignorance] little by little has carried life out onto the deep [sea] and stirred up from the bottom great waves of war" (5.1430–1435).

The connection between the development of the arts and war includes but is not restricted to the fact that many or most arts can be used in war. Ignorance of the limit of true pleasure causes people to seek ever new pleasures and new things; this search stimulates the development of the arts. Reciprocally, the development of the arts continually presents people with new things, new pleasures, new objects of desire, and thus feeds and inflames their unlimited desire for new things. The *unlimited* desire for things is the fundamental cause of both the development of the arts and war.[26]

As the arts develop, they make possible and contribute to the true human wisdom on which genuine happiness can rest. But the role they mainly play in most people's lives is not valuable; they are allied with, and serve, human ignorance and immoderation. Lucretius is not cheerful about the possibilities of political wisdom; he does not hold forth much hope that politics can provide genuinely moderating guidance for people's lives. Politics itself is too infused with irrational fears, false opinions, and vain desires. Real clarity about our human situation, moderate acceptance of our limits, and true peace of mind cannot be produced by political means. They can only result from an individual's own efforts in association with some good friends.

CONCLUSION

Today's technology differs so vastly in magnitude and scope from the arts of former times that it unquestionably poses problems not faced by ancient thinkers. The clearest example: ancient thinkers surely did not reflect on the human capacity to alter the world's climate or to befoul the seas. What, then, can one gain from understanding the thinking of the ancients on these matters? At a minimum, a firmer grasp of just what the modern technological enterprise involves. Its distinctive character, its basic views of human nature and the place of humankind in the world, its goals and the means to them can all become clearer to us against the background provided by a better understanding of alternative ancient views.

But perhaps something more is to be gained. Perhaps the ancients profoundly and correctly analyzed the fundamental problem in this

26. This discussion draws on a fuller treatment in James H. Nichols, Jr., *Epicurean Political Philosophy: The "De rerum natura" of Lucretius* (Ithaca: Cornell University Press, 1976), pp. 167–76.

domain: to set salutary, indeed indispensable, limits on the daring, the hopes, the deeds of this clever, artful human being. Just laws aiming to achieve the common good, reverence for what is higher than individual desires, and, above all, the moral virtue of *sophrosyne* (moderation, temperance, self-knowledge, and self-control): these may remain now, as they seemed to the ancients, the crucial means necessary for overall human guidance to good ends. Specific problems posed by technology require, of course, specific policies and even institutions appropriate to today's novel circumstances, but the basic moral necessities may remain such as the ancients found them to be.

To these reflections someone might well evoke the unprecedented magnitude of human power in relation to our natural world and proceed to ask: Does this not place unprecedented burdens of responsibility on us, heavier than those of the mythical Atlas? Must we not, as never before, answer for the continuing existence of a natural world hospital to life itself? Do we not therefore stand in need of some altogether new ethical guidance, or, as is often said, must we not create a new ethical paradigm?

Perhaps. But some critical reservations deserve a hearing. Powerful though modern technology has made us, is it not still presumptuous in the extreme to compare our power to that of the whole nature of things? If, as Lucretius believed, the universe is infinite, then, to quote him, "by the holy hearts of the gods, who pass a placid time and a serene life with tranquil peace!—Who has power to rule the sum of the immeasurable, who can hold in his controlling hand the strong reins of the deep?" (2.1093–6).

If you reject the infinity of the universe in favor of modern astronomical views, which hold that the sun's explosion must eventually burn the earth to a crisp, does not the idea of our responsibility for nature seem no less preposterous? These questions tend in no way to belittle the importance of our unprecedented need to plan our actions with intelligent consideration of their overall global effects; these questions seek rather to place that need in a less apocalyptic perspective as regards the whole of nature and our place in it.

Let me end with two words of caution concerning new ethical paradigms. If taken very seriously, a new ethical paradigm is a most weighty matter, of the greatest difficulty to bring about, and momentous in its consequences. After all, what experience of new ethical paradigms can we look to? I should suggest as an appropriate exam-

ple precisely that new way of thinking about ethics and political society introduced, in intimate connection with modern science and the
vision of modern technological society, by such thinkers as Francis
Bacon and René Descartes. This new ethical paradigm surely brought
about huge changes in our world and world view, leading to, among
other things, just those circumstances that cause some people today
to call for new ethical paradigms. (And these changes took place
despite the continued existence of many people—Christians, Aristotelians, perhaps even an occasional Epicurean—who held to older
ethical views, which, in fact, have never been wholly expunged from
our modern world.) Have people who voice this demand for new
ethical paradigms really thought through what is at stake? And have
we any sufficient grounds for believing that we know how to bring
about momentous desirable changes in human feelings, choices,
and actions?

To approach this same matter from a different angle, suppose that
moderation *(sophrosyne)*, as understood by the ancients, is the one
thing most necessary for setting limits on human conduct. This truth,
if truth it is, could provide us with some sound overall guidance for
dealing with the problems of technology, but no global solution. We
must still look at technological problems in all the relevant detail of
their circumstances; we still have to work at devising laws, policies,
institutions, incentives, and international agreements to grapple with
these problems; and we must still constantly struggle with how,
through education, to strengthen this undramatic old virtue of moderation, so long out of favor. Small wonder that one might turn instead to discourse on a new ethical paradigm. Does it not offer the
hope of a quick, intellectually radical solution for our problems? But
if this hope is utopian, it must be viewed as a distraction from harder,
slower, but potentially effective work that needs to be done.

Suggestions for Further Reading

Aeschylus. *Prometheus Bound*, trans. Herbert Weir Smyth. In *Aeschylus* I. Cam-
    bridge: Harvard University Press (Loeb Library), 1946.
Hesiod. *Works and Days* and *Theogony*, ed. and trans. Hugh G. Evelyn-White. In
    Hesiod, The Homeric Hymns and Homerica. Cambridge: Harvard University Press
    (Loeb Library), 1967.
Lucretius. *On the Nature of Things.* A good translation appears in *Titi Lucreti Cari*

Technology in Classical Thought   45

de rerum natura libri sex, ed. Cyril Bailey. More easily found is Lucretius, *On Nature,* trans. Russel M. Geer. Indianapolis: Bobbs-Merrill, 1965.

Plato. *Protagoras,* trans. W.R.M. Lamb. In *Plato: Laches, Protagoras, Meno, Euthydemus.* Cambridge: Harvard University Press (Loeb Library), 1962.

Sophocles. *Antigone,* trans. Elizabeth Wyckoff in *Sophocles I,* ed. David Grene and Richmond Lattimore. Chicago: University of Chicago Press, 1954.

Strauss, Leo. *The City and Man.* Chicago: Rand McNally, 1964.

GOSHEN COLLEGE LIBRARY
GOSHEN, INDIANA

2 Medieval Christianity: Its Inventiveness in Technology and Science

STANLEY L. JAKI

The word "medieval" is a derivative of *medium aevum* (Middle Ages), a term unknown to the medievals. Had anyone among their contemporaries tried to fasten that term upon them, they would have strongly protested. From the closing decades of the twelfth century, the latest possible point to mark the beginning of the High Middle Ages (a late nineteenth-century term), the medievals were very conscious of being in the forefront of progress, which is the very opposite of being caught somewhere in the middle between the old and the new. The idea of a progressive history was, in fact, formulated in the medieval School of Chartres.[1]

Those in that school and other twelfth-century scholars in the Latin West knew all too well the immense measure of their debts to ancient Greece and Rome. John of Salisbury, first a teacher in the School of Chartres and then bishop of Chartres, spoke of ancient Greece and

1. See M.-D. Chenu, *Nature, Man, and Society in the Twelfth Century: Essays on New Theological Perspectives in the Latin West,* with an introduction by Etienne Gilson; selected, ed., and trans. by J. Taylor and L. K. Little (Chicago: University of Chicago Press, 1968), pp. 166–67. See also R. W. Southern, "Aspects of the European Tradition of Historical Writing. 2. Hugo of St. Victor and the Idea of Historical Development," *Transactions of the Royal Historical Society* 21 (1971), 159–79.

Rome as a giant's two shoulders. It was only by standing on those shoulders, John of Salisbury warned, that he and his contemporaries saw farther than their remote predecessors.[2] For all his genuine modesty, John of Salisbury did not mean to disparage thereby his and his colleagues' conviction of seeing very much farther than the ancients.

And see they did many novel things, either by achieving them or by boldly conjuring them up as so many marvelous characteristics of a perhaps not too distant future. Toward the middle of the fourteenth century, which may mark the end of the High Middle Ages, the medievals coined the word "modern" and applied it to themselves. The word certainly contained the meaning of being modish, but this is precisely what progress is all too often. Even today, to progress is to come up with ever new modes, manners, fashions, and novelties— so many marks of an unrelenting readiness to innovate and invent.

The most monumental evidence of medieval inventiveness relates to architecture. The newly introduced buttresses and pointed arches allowed Gothic cathedrals to soar to daringly new heights. No less innovative were the techniques, many of them unknown today, that produced the enchanting hues in the Cathedral of Chartres and in many other medieval places of worship. Without taking a cue from the few traces of bright color remaining on Greek sculpture, the frieze of the Parthenon, for example, medievals displayed the same exuberance in painting the interior walls of many of their cathedrals.

Medievals were no less notably inventive in music, social organization, philosophy, and, last but not least, technology and science. Awareness of all this had for some time been close to the vanishing point when the expression *medium aevum* made its first appearance in the closing decades of the seventeenth century and inspired the coinage of terms such as "Mittelalter," "Moyen-Age," and "Middle Ages." Chief responsibility for this resides with the immensely popular textbooks on history by Christoph Keller, a German educator.[3] As one living in the latter half of what Alfred North Whitehead called the century of genius, Keller had no choice but to see a mere transitional period in what stretched between classical and modern times. The latter began, according to Keller, with the Reformation of faith

2. *The Metalogicon of John of Salisbury: A Twelfth-Century Defense of the Verbal and Logical Arts of the Trivium,* trans. with introduction and notes by D. D. McGarry (Berkeley: University of California Press, 1955), p. 167.

3. See G. Gordon, *"Medium aevum and the Middle Age"* (Oxford: Clarendon Press, 1925), pp. 4–28 (S. P. E. Tract No. 19).

and the Baconian instauration of knowledge. The Renaissance as a distinct age was yet to be invented by Jacob Burckhardt around 1850.

By then the word "medieval" had for some time been also a term of abuse and has remained so to the present time, a fact still to be recognized by makers of English dictionaries.[4] In actual use the word is all too often synonymous with such unflattering adjectives as "regressive," "obscurantist," and "barbarian." As a term of abuse or contempt, it shows up in the most unwarranted contexts. Undoubtedly, for the public relations man of a state-owned automobile factory in a Communist regime in 1990 it was catastrophic to find that on the day after New Year's none of the 1,500 workers showed up for the morning shift. It may, however, be doubted that this fact justified the official in question to look at the abandoned production lines and sigh: "It's like a medieval plague."[5] He should instead have spoken of "counterrevolution," the plague most feared and resisted in Marxist realms, but he no longer had this privilege. His workers simply extended their celebration of New Year's Day beyond the legal limit by sensing a "medieval" straw in the wind. The still partly Communist regime in Prague had, by giving amnesty to all political prisoners, resorted to the typically "medieval" gesture of forgiving in the uncanny expectation that within a few months they themselves might need a similar gesture, and in an ample measure at that.

But to return to the middle of the nineteenth century. By then the period known as Romanticism had run its course after having restored an appreciation of Gothic architecture, which had been held in contempt for many centuries. Auguste Comte, the father of positivism, even made it fashionable to speak appreciatively of medieval social organization. In the second half of the nineteenth century something more began to be seen in Scholastic philosophy than a slavish imitation of old phrases and a bent for logic chopping. Still later, by the mid-twentieth century, leading American universities vied with one another in setting up chairs and courses in medieval philosophy. No less eagerly they began to find a place in their libraries for books and periodicals on this formerly suspect subject.

But even then it was not yet academically respectable to speak of medieval inventiveness in science and technology. It was one thing to decry as dinosauric a book on the history of astronomy in which

4. The omission is certainly glaring in the multivolume *Oxford Dictionary of the English Language* (1933), its Supplement volumes, and massively increased 2d ed. (1989).

5. Quoted in *New York Times*, January 4, 1990, p. A15.

the medieval period was represented by three or four literally empty pages.[6] It was another, and a rather risky affair, to speak of medieval science with a touch of genuine admiration. This was all the more curious because by the 1930s and 1940s a massive scholarly evidence about science in the Middle Ages had for some time been available. Even when, from the 1960s on, it could appear unscholarly to ignore medieval science, appreciative references to it were often qualified with remarks that amounted to the art of damning with faint praise.

If the word "damning" is taken in its erstwhile theological connotation, the clue to this curious performance easily comes into focus. "Medieval" and "Middle Ages" became terms of abuse during the Enlightenment, an age that saw its chief task as replacing the darkness of faith with the light of reason. The faith was, of course, Christianity, with its insistence not only on a personal Creator but also on a supernatural revelation. That insistence found its most notable social embodiment in medieval Latin Christianity. As to reason, the chief connotation given to it by the champions of the Enlightenment related to Newtonian science. In England Alexander Pope declared that once God "let Newton be, all was light." In France Voltaire's career as the chief of *philosophes* started with his popularizations of Newton's physics, about which he knew no more than popular accounts. In Germany Kant wrote in the preface to the second edition of the *Critique of Pure Reason* that "when Galileo caused balls to roll down an inclined plane, . . . a light broke upon all students of nature."[7]

In writing the *Critique* Kant meant to spread that light in the deepest possible sense. He believed he was offering the only valid and rationally respectable form of metaphysics that, unlike traditional or Scholastic metaphysics, had man instead of God as the ultimate source of reason. Moreover, the antitheistic arguments of the *Critique* had, according to Kant, that very final validity which in his eyes was the hallmark of Newtonian physics.

Kant had but a smattering of knowledge of Newton's physics.[8] Were

6. H. S. Williams, *The Great Astronomers* (New York: Newton, 1932), pp. 97–99.

7. *Immanuel Kant's "Critique of Pure Reason,"* trans. N. K. Smith (New York: St. Martin's Press, 1965), p. 20.

8. No serious grasp of Newton's physics can be detected in the writings of M. Knutzen, who, according to a still well-entrenched academic cliché, is supposed to have introduced the young Kant into that physics in private instruction. See my introduction to my translation with notes of Immanuel Kant, *Universal Natural History and Theory of the Heavens* (Edinburgh: Scottish Academic Press, 1981), pp. 28–29.

this to be widely perceived, broadly shared doubts might arise about the reliability of Kant's attack on the cosmological argument, which is man's principal intellectual avenue to the recognition of God. More likely, modern Western man will find ever fresh pretexts to safeguard the absolute autonomy bequeathed to him by the Enlightenment. The latter is but a continuation of the Renaissance, which began with conscious reaching back to the pantheism and all too often crude paganism of classical antiquity.

We are still in that Renaissance, and its end is far from being in view. That Western man is today a Renaissance man is conveyed by his boasting of his now living in a post-Christian age. The deity of this age is science, in which modern Western man finds the chief guarantee of his conviction of being his own absolute master and accountable to nothing higher. No wonder that modern Western man resists few things more violently than the claim, however scholarly, that the Latin medieval centuries, so synonymous with medieval Christianity, were distinctly and, in fact, creatively innovative in technology and science. This and nothing else is the true explanation for a reluctance, now almost a century old, to accept without major reservations Pierre Duhem's most scholarly unfolding of massive evidence about science in the Middle Ages.[9]

Duhem and a few other Catholic historians of science undoubtedly saw in medieval science a credit to Catholic faith. Duhem himself viewed science in the Middle Ages as a proof of Christ's promise that those who seek first the Kingdom of Heaven will reap benefits on earth as well.[10] Duhem would have been the last to assert that they would be the sole recipients of such benefits. As a great logician, he would readily have seen through the apparent evenhandedness of the statement: "To deny or minimize the culturally desirable products of Christianity because one may consider Christianity to be a fabric of illusions, is as subjective as to argue that Christianity is true because some of its effects seem salutary."[11]

9. For details, see chap. 10, "The Historian," in my *Uneasy Genius: The Life and Work of Pierre Duhem* (Dordrecht: Martinus Nijhoff, 1984).

10. See his letter of May 21, 1911, to P. Bulliot, translated in full into English in my *Scientist and Catholic: Pierre Duhem* (Front Royal, Va.: Christendom Press, 1991), pp. 235–40, from Hélène Pierre-Duhem, *Un savant français: Pierre Duhem* (Paris: Plon, 1936), pp. 158–69.

11. Lynn White, Jr., "What Accelerated Technological Progress in the Western Middle Ages," in A. C. Crombie, *Scientific Change: Historical Studies in the Intellectual, Social, and Technical Conditions for Scientific Discovery and Technical Invention from Antiquity to the Present. Symposium on the History of Science. University of Oxford, 9–15 July 1961* (New York: Basic Books, 1963), p. 291.

Coming as it does from the late Lynn White, Jr., a prominent histo-rian of medieval technology, the statement should seem most unbal-anced for a historical reason. It relates not to the Age of Faith but to the Age of Reason or to modern times in general. Ever since Voltaire, Christians, or Catholics in particular, have learned to live with the cultural shibboleth that their faith and science are mutually exclusive realms. Even today, three-quarters of a century after Duhem, they do not seem overly eager to make much of his findings. Yet they could learn nothing less than that they are under a make-believe siege based on their and their antagonists' ignorance about what has taken place in the history of science. But only the dullest Catholic intellectual would fail to sense the plight of modern unbelievers confronted with some facts of medieval science. Those who staked their unbelief or agnosticism on the rationalistic account of the roots of science will hardly learn to live with their scholarly uprooting.

So much for the broader ideological stakes behind the topic of this essay—stakes that ought to be spelled out at the very outset as a service to intellectual honesty. Such honesty may prompt some read-ers to admit, at least to themselves, that their boasting about not having a religion may be a mere camouflage for their avid espousal of the religion known as secularism. At any rate, it should be clear that one can no more appraise the Middle Ages in general, or its science in particular, while disregarding the Catholic creed than one can understand the formation of the United States without studying the Federalist Papers or understand England apart from the Magna Carta. This is even true about medieval inventiveness in technology, a field apparently unrelated to theology. But the theological roots of medieval science will appear crucially decisive with respect to the most significant scientific breakthrough in the history of science, which was achieved during the High Middle Ages. It was, as will be seen, a breakthrough to the very possibility of science insofar as it deals with this very physical world, a world of things in motion.

In view of the secularist tone of most discourse about science, it becomes almost obligatory to offer a word of apology whenever those medieval centuries are claimed to be progressive in the scientific and technological sense in which progress is almost exclusively under-stood nowadays. The editor of *Smithsonian* Magazine seemed to be very much on the defensive when he introduced an article by Lynn White, Jr., with the caption: "Middle Ages weren't 'dark' since inven-tion found expression in architecture, mills, roads—and ideas bearing fruit in the 20th century."[12] Most likely the editor rather than White was responsible for the title of his essay, "Medieval Europe Foresaw Planes, Cars, Submarines."

This is certainly a catchy title but hardly a promising catch. For if only visionary cars, planes, and submarines can be caught in the net of the student of medieval technology, the efforts are largely wasted. Yet the catch should seem very tangible and exciting if White was right in stating almost at the outset: "While nothing of ancient Greek or Roman engineering perished along the road, the attitudes, motivations and most of the basic skills of modern technology before the electronics revolution originated not in Mediterranean antiquity but during the 'barbarian' Middle Ages."[13]

White was not alone in making this startling statement. In 1976 the MIT Press published Arnold Pacey's study on the ideas and idealism in the development of technology, which contains a long chapter titled "A Century of Invention: 1250–1350."[14] Those hundred years are certainly a part of the High Middle Ages. Moreover, in Pacey's book only that chapter has the word "invention" in its title, although eight centuries prior to 1900 are covered in its eight chapters. Clearly, the High Middle Ages had to be particularly inventive in matters technological.

The 1970s also saw the publication of *Turning Points in Western Technology* by D. S. L. Cardwell, who specifies four stages in the development of technology. The last stage began in the late nineteenth century with the establishment of research laboratories. This was preceded by the Industrial Revolution, which in turn owed much to the second stage, or the publicity given to mechanical arts by Bacon and other seventeenth-century empiricists. Cardwell locates the first stage in the Middle Ages and refers to the stationary character of Greco-Roman technology. Furthermore, he describes the advance over classical antiquity made by medieval engineers as something amounting to a mutation. Overused as this word may be among academics today, it should at least alert them to the magnitude of the change:

12. *Smithsonian*, October 1978, 114–23.

13. Ibid., p. 115. This is a point also emphasized by White in his essay cited in note 11 (p. 273), and previously in his "Technology and Invention in the Middle Ages," *Speculum* 15 (1940), 150.

14. Arnold Pacey, *The Maze of Ingenuity: Ideas and Idealism in the Development of Technology* (1974; repr. Cambridge: MIT Press, 1976).

Such, in fact, was the precocity, diversity and importance of medieval technics that one is inclined to believe that a *mutation*, philosophical or spiritual, had occurred, so that medieval civilisation was sharply and basically different from all those that had preceded it. The technics of the period were precocious in that they often ran far ahead of the very limited scientific knowledge available. They were diverse in the vast range of activities they were concerned with and they were important, not merely by virtue of their cumulative effects on the economic development of society but also because several medieval inventions decisively and fundamentally changed man's outlook on the world. Very few subsequent inventions had the same universal significance in this respect as did the weight-driven clock and the printing press. Modern society is still basically conditioned by these two medieval inventions.[15] (Emphasis added).

Such glowing appraisals of medieval technology represent a new point of view, perhaps a budding consensus. It certainly contrasts with the view that had set the standard until the 1950s. A memorable instance of that view can be found in a five-volume history of technology produced with the collaboration of many specialists under the editorship of Charles Singer, a leading historian of science half a century ago. In the epilogue to the second volume dealing with medieval technology Singer presented the Middle Ages as the end of a very long preparatory period, possibly even less than a transition.

The chief characteristic of that period, according to Singer, was the flow of technological know-how from the East to the West. Only when the Middle Ages ended, Singer wrote, "had the east almost ceased to give techniques and ideas to the west and ever since has been receiving them." According to Singer, this strange reversal should prompt two reactions. First, one has to marvel at the astonishingly effective accumulation of technological skill even in the absence of scientific guidance. Second, one has to be deeply impressed by what he called a "sharp contrast."[16] It consisted in the difference between a technological development that for many thousands of years did not depend on the guidance of science and a postmedieval development in which dependence on science has been essential for technology.

15. D. S. L. Cardwell, *Turning Points in Western Technology* (New York: Science History Publications, 1972), pp. 211–12.
16. *A History of Technology*, ed. Charles Singer et al., vol. 2: *The Mediterranean Civilizations and the Middle Ages* (New York: Oxford University Press, 1956), p. 774.

Clearly, one is in the presence of two very contrasting appraisals. In the one, voiced by Singer, the Middle Ages are at best a recipient; in the other, voiced by White and others, the medieval centuries are a creative starting point to which our modern times, so technological and scientific, are heavily indebted. The two viewpoints differ even in the counting of medieval centuries. Did the Middle Ages begin around 400 and end shortly after 1300? In this view, which is essentially that of Singer, the Middle Ages consist of seven centuries of stagnation followed by two centuries, the twelfth and thirteenth, of some dynamism, tempered if not hampered by the flowering of Scholasticism. Or did the Middle Ages begin with Charlemagne, about 800, and end about 1450, or just before the great majority of Plato's writings became known in the West?

The latter view may be preferable for two reasons. One is that in this case the end of the Middle Ages is closely followed by the beginning of the Renaissance in its deepest aspect, best revealed by the humanists' enthusiastic response to the sudden availability of most of Plato's works.[17] Underlying that enthusiasm was a conscious reassertion of the ideals of Greco-Roman humanism, a humanism very different from the medieval concept of man's nature and purpose. Also, if one begins the Middle Ages with the Carolingian renaissance of learning, one is spared from seeing cultural unity in the four centuries postdating the formal demise of the Roman Empire in 410. Those four hundred years were an epoch of disintegration, upheavals, fragmentation, and instability—so many evidences not so much of a culture as of a transitional period between cultures. To be sure, the church was there as the only stable point, but its culture-making ability still had to make itself felt. The latter is a question not only of faith but of sociological conditions as well.

The chronological definition of the Middle Ages as an epoch from 800 to 1450 provides a logical ground for the historian insofar as the work of evaluation has to be concerned with comparison. Since such a comparison is usually a reference to the preceding culture, the

17. The immediate effect was that Ficino was instructed by his princely patron in Florence to give priority to the translation of Plato's works into Latin. Efforts followed to give intellectual respectability to the idea of eternal recurrence, the quintessence of classical paganism and, for Christians, its most unacceptable aspect. For details, see chaps. 8 and 11 of my *Science and Creation: From Eternal Cycles to an Oscillating Universe* (1974; 2d enlarged ed., Edinburgh: Scottish Academic Press, 1986; repr. Lanham, Md.: University Press of America, 1990).

evaluation of the Middle Ages has to be done with an eye on the Roman Empire.

During both the Middle Ages and Roman times material and organizational needs were keenly felt. The Roman Empire had periods of famine, as did the Middle Ages. Both heavily depended on agriculture and relied on manual labor. One was largely served by slaves, the other by serfs. Although the Christian serfs received more humane treatment than did the Roman slaves, the physical conditions of their lives were not much different. At any rate, the pagan Roman landowner had the same interest in a good harvest as did the medieval feudal lord. Why is it then that a major improvement of the plow had to wait for the Middle Ages? Unlike the traditional Roman plow, the medieval plow had, in addition to a vertical knife, a horizontal shear to slice under the sod and a moldboard to turn it over.[18]

Material needs, of course, must have played a part in that very important technological invention, but why then did those needs fail to spur the inventiveness of the Romans of old? They and the Greeks and other ancient nations that made much use of horses in warfare failed to exploit an elementary fact about the horse's anatomy. If a horse is made to pull something by means of a strap around its neck, it not only does so ineffectively but also risks being exposed to strangulation. However, the weight-pulling effectiveness of the horse can be increased fourfold with a breast harness, a medieval invention, as is the nailed horseshoe and the harnessing of horses in front of one another.[19] Their use of plowing the fields is first depicted on the Bayeux tapestry, more famous for showing Halley's comet.

Those medieval inventions made an agricultural revolution. Together with the introduction of three-field crop rotation, they almost doubled the harvest. While the physical conditions of medieval serfs remained much the same as those of Roman slaves, they ate much better. A further proof of this is the much wider use of watermills in the Middle Ages compared with Roman times. In England, for instance, a late eleventh-century count of three thousand villages registered almost twice as many watermills, a situation not untypical of other parts of Latin Christendom.

Watermills were needed for more than the grinding of grain. Fullers found a revolutionary use for them, proof of the fact that the clothes worn by medieval people were of better quality than those of their

18. See White, "Technology and Invention in the Middle Ages," pp. 153–54.
19. Ibid.

Roman forebears. (Those clothes also had buttons and pockets, two small but most useful advances achieved in the art of dressmaking during the Middle Ages.) The use of watermills by fullers is tellingly evidenced in the description of the abbey of Clairvaux, given around 1180 by a Cistercian monk.[20] The description is an encomium to machinery insofar as it relieves man of hard labor.

In the description of the abbey the visitor approaching it along the river first sees two hills, a vineyard and an orchard respectively, between which nestles the abbey itself. On the near side of the orchard the visitor finds a fishery, made possible by damming the river. From there the river is channeled into the abbey, but in such a way that in case of flood its surplus water would flow around the abbey along an auxiliary canal. From here on, let the anonymous monk take over:

> Once the river is let inside the abbey through a sluice, it first rushes against the flour mill, where it is very solicitous and occupies itself in many tasks, both in grinding the grain and in separating the flour from the bran. . . . But the river is far from being through with its work. It is invited by the fullers who labor next to the flour mills and who rightly demand that just as the river was busy in the mills so that the brethren may be fed, it should also assist the fullers so that the brethren may also be clothed. The river does not decline any work the fullers demand from it. By alternately raising and lowering the heavy stumps, or pestles, or if you wish hammers, or wooden legs (because this name better fits the fullers' jumping style of action), the river relieves them of their really heavy work. Or, if one may inject into a serious topic a remark in a light vein, the river takes upon itself the fullers' punishment for their sins. O God, in your goodness you provide so much respite to your poor servants lest they be overtaken by much sadness! How much relief do you administer to penitents from their sins, lest they feel oppressed by the hardship of their work! The backs of how many horses would be broken and how many men's arms greatly exhausted by a labor from which this river gratuitously dispenses us and without which no clothing, no food would be prepared! The river therefore shares our life while demanding nothing in a way of recompense for its work, which it performs under the blazing sun, except that it may freely leave after it has diligently done its

---

20. My translation is from the Latin text, "Descriptio positionis seu situationis monasterii Clarae-vallensis," in J.-P. Migne, ed., *Patrologiae cursus completus. Series latina* (Paris: J.-P. Migne, 1844–64), vol. 185, cols. 569–74.

work. It has turned so many big wheels in rapid rotation that it emits froth so that it looks tamed and increasingly so as it moves on.

Instead of following the further course of the river through the numerous other workshops, such as the tannery and laundry, which are also enthusiastically described by that anonymous monk, let us return to his reference to the fullers' up-and-down legwork, now performed by trip-hammers. As well as describing the river's turning "so many big wheels in rapid rotation," he attests to the use of one of the foremost medieval technological inventions, the cam. It does nothing less than transform rotary motion into linear thrust, and vice versa. The medieval introduction of mechanical saws, not mentioned by the monk, is another example of the use of cams.

When in 1444 Bessarion, the future cardinal, arrived in Italy from the Peloponnesus, one of the things that struck him in particular was the widespread use of mechanical saws. He quickly urged his countrymen to send young mechanics to Italy to learn about this and other labor-saving machinery.[21] Here is a long-ignored witness to the technological superiority of Western over Eastern Christendom, a superiority also true with respect to the Muslim world.[22]

By the time of Bessarion's visit, trip-hammers had been in use for over four centuries in Latin Christendom, making possible not only the increased production of felt from raw cloth but also the more effective breaking up of various mineral rocks. One of the results was a truly transparent glass that made possible the invention of eyeglasses about 1280. This feat is still widely attributed to the Renaissance, although documentary evidence has been readily available in scholarly journals for more than half a century. Clearly, one is entitled to ask the question, "How is it that the importance of the late thirteenth-century Italian invention of spectacles has not been more generally appreciated?"[23] Academics, often bespectacled, would hardly

---

21. For quotations from Bessarion's letter, see A. G. Keller, "A Byzantine Admirer of 'Western' Progress: Cardinal Bessarion," *Cambridge Historical Review* 11 (1954–55), 343–48.

22. As can easily be seen from a perusal especially of chap. 6 in the fourteenth-century Muslim cultural historian Ibn Khaldoun's much-praised *Muqqadimah* (that is, the preface and bk. 1 of his *Universal History*). That chapter is available almost in full in the one-volume abridgment published by N. J. Dawood (Princeton University Press, 1969) from F. Rosenthal's three-volume translation. For further discussion, see my essay "The Physics of Impetus and the Impetus of the Koran" (1984), reprinted in my *Absolute beneath the Relative and Other Essays* (Lanham, Md.: University Press of America, 1988), pp. 140–52.

23. A question of Lynn White in his "Technology and Invention in the Middle Ages," pp. 142–43.

disagree about the crucial cultural importance of eyeglasses. Then why are many of them disinclined to recognize the medieval provenance of that discovery? Perhaps because they would then be forced to note the specter of Christian theology and react to it with King Lear's words: "That way lies madness."

Modern Western academics are certainly reluctant to spread the word that medieval trip-hammers were responsible for the making of paper in large quantities. They are even less ready to consider that the greater availability of paper, as distinct from the scarce and very expensive parchment, made it natural to think about paper as a means for the wider distribution of written texts. Herein may lie the source of the medieval invention of printing. First came the extensive use of block printing, in itself a borrowing from the East, and then the introduction of movable type. Whether the idea of movable type came to the medievals from the East is a moot question.[24] At any rate, unlike the medieval Chinese and Koreans, the Christian Latin medievals made the most of printing from the early 1400s on. The Gutenberg Bibles would seem to be as impossible an achievement on a first try as would a Boeing 707 for the Wright brothers' first flight or a Mustang for Henry Ford's first car. Gutenberg's contribution to the art of printing may not, in fact, have exceeded "a successful method for printing initial letters in color." This statement, however startling, is now a half century old and comes from Pierce Butler's *Origin of Printing in Europe*. Although published in 1940 by the University of Chicago Press, the book failed to make a dent in cultural consciousness despite its revolutionary message.[25]

Scholarly research notwithstanding, the invention of printing, which may have been an indirect outgrowth of the medievals' interest in watermills, is still largely credited to Gutenberg and the Renaissance. Not so completely ignored is the medievals' inventiveness in constructing windmills. They appeared everywhere in Europe during the latter part of the twelfth century, with a horizontal axis that had

24. See ibid., p. 146, for bibliography and discussion.

25. Pierce Butler, *The Origin of Printing in Europe* (Chicago: University of Chicago Press, 1940), p. 143. Butler further decries Gutenberg's fame as the inventor of printing as "the purest example of folklore ever developed in modern times." This folklore was given further wide currency in D. J. Boorstin, *The Discoverers* (New York: Random House, 1983), a book that, though written by a former director of the Library of Congress with an army of research assistants at his disposal, embodies an almost thematic blindness about the technological and scientific inventiveness of the Middle Ages, while presenting the medievals as being blind prisoners of Christian dogma.

been unknown in the East. Furthermore, the vertical plane of the rotation of the blades was slightly tilted in order to obtain a better torque. This invention, too, was in part motivated by material needs, but other motivations must have been present. Otherwise it is difficult to explain the neglect of windmills in the Far East, especially in areas with strong winds, such as Tibet. There they largely drove prayer wheels. Since the medievals also prayed, and certainly their monks did, the ideology underlying their prayers must have had ingredients not possessed by the prayers of Buddhist monks.

Again, if practical need is the mother of all inventions, one is hard put to account for the primitiveness of Roman road building. To keep together their vast empire, the Romans certainly needed good and easily reparable roads. They invariably used massive stone slabs, but water readily seeped between the big slabs and destroyed their alignment. Although the medievals found that it was better to leave the open roads unpaved, they were the first to cover city roads with square stones.

Mere material need is hardly the explanation for the stunning architectural innovations of the Middle Ages. To be sure, something practical had to be done to the old Roman-style buildings whose flat, timber-supported, wood-paneled roofs and open fireplaces made them so many fire hazards. One remedy was to construct mantel-type fireplaces with chimneys; another was to replace the flat roofs with masonry vaulting. The Gothic-style church represented a saving in stones because the pointed arches did not require the support of walls as thick as those needed for the cylindrical roof of a Romanesque church. No material need can be seen for the raising of Gothic churches to ever greater heights, a feat made possible by the invention of side buttresses. The additionally felt need may have been a satisfaction of vainglory, but was such a need not rather an occasional deformation of genuine spiritual longing? While the predominantly horizontal lines of Greco-Roman temples symbolized a nature-bound religious perspective, Gothic spires symbolized the upward reach of a distinctly supernatural vision.

Again, it would be very difficult to specify the material need that motivated efforts to which Robert the Englishman referred in 1271. According to him, "clockmakers are trying to make a wheel that will move exactly as the equinoctial circle."[26] What he meant was that

26. The first to publish that statement was Lynn Thorndike, "The Invention of the Mechanical Clock," *Speculum* 16 (1941), 242–43.

various craftsmen were trying to invent a mechanism that would make possible a weight-driven clock making a full turn every twenty-four hours. The pivotal part of such a mechanism was a double-feedback device whereby the gravitational pull of a weight was made to act at regular intervals, preferably measured in seconds or half seconds.

The periodic rewinding of such a clock, to say nothing of its construction and maintenance, could at first be just as cumbersome as the periodic refilling of water clocks and the turning over of sand clocks. Also, the steady pull of weight-driven clocks was too weak to be useful for industrial purposes. Nor could the need for such a clock relate to better timekeeping. Available sand clocks and water clocks were exact to within a couple of minutes over a period of several hours. Greater precision in marking time was hardly needed in view of the slowness and unreliability of transportation. There is little substance to Lewis Mumford's often-quoted statement that the invention of weight-driven clocks was motivated by an anticipation of the modern strictly regulated daily work schedule.[27]

The medievals were, of course, ready to turn their mechanical clocks to practical uses. This is shown by the oldest reference to mechanical clocks, a regulation issued in Sarum (Salisbury) that forbade the purchasing or selling of "flesh, fish or other victuals before the clock of the Cathedral had struck one."[28] More important is the possible identity of that clock as the one in the left nave of Salisbury Cathedral, the oldest of early mechanical clocks extant, dating from 1386. Being a fairly elaborate mechanism, it suggests that cathedral clocks were numerous from the mid-fourteenth century on. The same period marks also the beginning of frequent references to the planetary system, or the system of the world, as a clockwork mechanism. As they celebrated the universe as a clockwork, Newton and Voltaire perhaps unknowingly repeated an expression that, though Greco-Roman in origin, first achieved popularity in the Middle Ages.

Much less did Newton suspect that a most notable medieval reference to the universe as a clockwork occurred in a context that contained, in substance at least, his first law of motion. While Newton knew that Descartes certainly preceded him in stating that first law,

27. Lewis Mumford, *Technics and Civilization* (New York: Harcourt, Brace and World, 1934), pp. 13–14.

28. C. F. C. Beeson, *English Church Clocks, 1250–1850* (London: Antiquarian Horological Society, 1971), p. 16.

the law of inertial motion, he wished posterity to remember him as its discoverer. To assure this, he spent precious time in his old age erasing from his notebooks references to Descartes. Not that Descartes was more ready to give credit to others, let alone to medievals. Though familiar with Scholastic thought as it was discussed in Jesuit schools around 1600, Descartes kept silent about his own indebtedness to the medieval tradition on an all-important point, the idea of inertial motion in the form of the theory of impetus.

The impetus theory was widely discussed in books written by teachers at such principal seats of late sixteenth-century Scholastic learning as the universities of Salamanca and Coimbra and the Collegio Romano. Earlier, and most important, Copernicus used the impetus theory in his *De revolutionibus* to answer objections created by the twofold motion of the earth.[29] Copernicus did not name his sources, nor did he claim originality. In all likelihood he had learned about the impetus theory while a student at the University of Cracow. Its library has several manuscript copies of Buridan's and Oresme's commentaries on Aristotle's *On the Heavens,* in which they set forth their theory of impetus between 1330 and 1380. Their commentaries were widely copied and carried to all major universities of Europe. By the time Copernicus came on the scene about 1500, the impetus theory had been a widely shared idea for a century or so, though not yet through printed books. As a student, Copernicus learned about the impetus theory from codices or from plain oral tradition.

Copernicus was a student at the time the practice of publishing turned from the reprinting of medieval manuscripts, often selected at random, to the printing of original texts. Manuscripts, among them Buridan's and Oresme's commentaries, which were not printed until the early years of the sixteenth century, suddenly fell into a total oblivion. They began to be thrown out or to gather dust in many places of learning including such prestigious ones as the Bibliothèque Royale of Paris (which later became the Bibliothèque Nationale).

Voltaire, who could not have given higher praise to the age of Newton, might have refrained from taking jabs at the Middle Ages had he suspected that Newton's first law was in substance anticipated in the medieval Sorbonne. Early twentieth-century and strongly Voltairian France gasped in disbelief when, between 1905 and 1913, Pierre

29. A point widely publicized in chap. 1, "The Conservatism of Copernicus," in H. Butterfield, *The Origins of Modern Science 1300–1800* (1949; new ed. 1957; New York: Macmillan, 1960).

Duhem unearthed the startling evidence.[30] It included the fact that both Buridan and Oresme based their impetus theory on the very foundation of the Christian creed, or the creation out of nothing and in time. Both were teachers in a Sorbonne that was a bastion of Catholic orthodoxy.

The capital texts of Buridan and Oresme are fully available in English in Marshall Clagett's *Science of Mechanics in the Middle Ages*[31] and in Edward Grant's *Sourcebook on Medieval Science*.[32] But the comments offered by them do not emphasize two most important points. One relates to Buridan's substantial anticipation of Newton's first law. Of course, Newton and Descartes had in mind a rectilinear inertial motion, while for Buridan (and for Copernicus and Galileo) the inertial motion of the planets and stars was still circular. Hence the insistence of Alexandre Koyré that the origin of modern physics, which is in a sense the origin of the idea of inertial motion, should be located not in the Middle Ages but in the seventeenth-century scientific revolution.[33] But is Buridan's notion of inertia as un-Galilean as Koyré wanted us to believe? Are Koyré's *Galilean Studies* a convincing refutation of Duhem's principal claims?[34]

The answers to these questions depend on an in-depth look at what lies beneath Aristotle's thinking about motion. For him motion always implies, in one way or another, that the moved body should remain in contact with the mover. The way can, of course, be almost "spiritual," such as the perennial desire of the celestial sphere to be moved by the Prime Mover. The latter, let it be recalled, is most likely a still "higher" celestial sphere, perhaps the empyrean heaven, but certainly not a Creator.

30. For details and documentation, see my *Uneasy Genius*, which contains a full list of Duhem's publications. The disbelief turned into resistance. Volumes 6–10 of Duhem's *Le Système du monde* were published with a delay of almost forty years and only through the heroic persistence of his daughter and only child, Hélène. This shocking academic scandal is told in full, on the basis of previously unavailable epistolary evidence, in my *Reluctant Heroine: The Life and Work of Hélène Duhem* (Edinburgh: Scottish Academic Press; Front Royal, Va.: Christendom Press, 1992).

31. *The Science of Mechanics in the Middle Ages* (Madison: University of Wisconsin Press, 1961), chap. 8.

32. *Sourcebook on Medieval Science* (Cambridge: Harvard University Press, 1974), pp. 275–80.

33. Most extensively in his *La révolution astronomique* (Paris: Hermann, 1971).

34. Originally published in three fascicles between 1937 and 1939. Had those three studies been graced with a name index when published as a single volume (Paris: Hermann, 1966), the far more frequent appearance there of Duhem's name than of any other would have immediately revealed Koyré's chief intention.

Less "spiritual," and indeed subtly physical or "ethereal," is the continuous contact whereby, according to Aristotle, the motion of the sphere of the fixed stars is transmitted to the planets. In discussing the cause of fully physical motions that take place below the moon's orbit, Aristotle reported the idea of *antiperistasis,* according to which a projectile keeps moving because the air that it separates closes in behind it and pushes the projectile forward.

What could Aristotle's reasons have been, one may ask, for not criticizing this rather absurd idea? It is here that an in-depth look is needed, and one that is unabashedly theological. Aristotle reasoned, of course, within the terms of a pantheistic religion for which all motions in the universe must not only be interconnected but be so in a very special manner: everything had to be in contact with everything else, with no real beginning anywhere. Therefore the notion of antiperistasis had to appear to Aristotle as something germane to his general idea of motion as having its source in the continuous contact of the moved with the mover. This idea rested ultimately on his belief that the universe was uncreated and forever in motion.

It was precisely in this distinctly theological point that Buridan's break with Aristotle lay. For Buridan, the medieval Christian, the world was created in time, as explicitly defined at Lateran IV in 1215. In other words, for him the past history of the universe, or of motion for that matter, was strictly finite. Moreover, unlike Aristotle's Prime Mover, the Creator had to be essentially distinct from the universe. As a result, the created universe had to have an autonomy, an autonomy not necessary but contingent on the Creator's choice. It is by that choice that the Creator gave reality to one out of an infinite number of possible universes. Consequently, the celestial sphere and the planets move, according to Buridan, not through a quasi-physical contact with the Creator, who in that case would degenerate into Aristotle's Prime Mover, but rather through an impetus (quantity of motion or momentum) conferred on them in the very moment of creation. Buridan adds the all-important point that the celestial bodies retain their impetus undiminished because their motion takes place in a frictionless region of the universe.

So much on behalf of Buridan's substantial anticipation of Newton's first law of motion. It would seem to contradict Koyré and his many disciples and followers. For them the great breakthrough to modern science was the shift from circular to linear inertia. Rather, the really substantial shift was the one from motion through perennial continuous contact to a motion through an initial impetus imparted in an instant. If the latter shift had not taken place, the shift from circular to linear inertia would have remained inconceivable. In other words, insofar as science is a quantitative study of things in motion and the first law of Newton is the basis of his other laws, one may indeed speak of the substantially medieval origin of modern science.

Moreover, that medieval origin is also of Christian origin in view of the impact on Buridan's thinking of the Christian dogma of creation out of nothing and in time, the second point not emphasized by Clagett and Grant. Buridan was part of a vast intellectual milieu that consciously relied on that dogma. This was an indispensable condition for the subsequent popularity of the impetus theory in the Christian medieval West. The importance of this point can best be appraised with an eye on medieval Jewish and Muslim milieus.

Surprising as it may seem, the doctrine of creation out of nothing and in time was much less emphasized in Jewish than in Christian ambience. For instance, in all of the many references to creation by Philo of Alexandria, only one constitutes an unambiguous endorsement of creation out of nothing.[35] Then as today, liberal Jewish intellectuals displayed a curious sympathy for what is best described as Spinozean pantheism.[36] Within orthodox Judaism it became a trend to avoid questions that are not explicitly posed in the Hebrew Bible.[37] There one would look in vain for an explicit statement of creation out of nothing.

Orthodox Muslim thinkers, such as al-Ashari and al-Ghazzali, turned the voluntarist presentation of God in the Koran into an early form of occasionalism. They frowned on the idea of natural law as a constraint on Allah's ability to do whatever he wants to do. Today's Muslim fundamentalism may indeed be a last-ditch stand against the encroachment of natural law on Muslim thinking through the industrialization of Muslim lands.[38] Not a few intellectuals in Muslim countries are now drifting into a pantheistic stance just as readily as

35. As rightly pointed out by F. V. Cournin, "Philo Judaeus and the Concept of Creation," *New Scholasticism* 15 (1941), 46–58.

36. A recent example is Yirmiahu Yovel's book *Spinoza and Other Heretics* (Princeton, N.J.: Princeton University Press, 1989), which received wide publicity in the United States through Flora Lewis's article "Israel and Relevance," *New York Times,* December 27, 1989, p. A23.

37. See my *Savior of Science* (Washington, D.C.: Regnery-Gateway, 1988), p. 70.

38. See my essay "The Physics of Impetus and the Impetus of the Koran."

Avicenna and Averroes espoused Aristotelian necessitarianism. In both cases a resort to concealment also proved to be necessary.

The effectiveness with which pantheism has always been resisted within Christianity has much to do with the dogma of Incarnation, a tenet wholly alien to Jewish and Muslim thinking. The dogma of Incarnation, or the doctrine about the *monogenes* or *only begotten* Son who is Christ, could not fail to serve as a strong barrier against pantheism.[39] Within the Greco-Roman form of pantheism, the universe was the "monogenes" or "only begotten" emanation from a divine principle not really different from the universe itself. John's references to Christ as the "only begotten" therefore constituted an enormous semantic and doctrinal break with classical antiquity. Historians of science still have to discover that break as the means to a better understanding of the medieval origin of science.

That break was ultimately responsible for another aspect of the medieval stirrings of modern science. Because belief in the Word-made-flesh became a widely shared climate of thought in the Middle Ages, effective resistance could be made against Cathars and kindred sects. It passes nowadays for scholarship to present Cathars as innocent victims of a power-hungry intolerant church.[40] The scholarship in question is very shortsighted with respect to the fate and fortunes of science. The medieval stirrings in science would have been nipped in the bud if the hatred for matter cultivated by Cathars and related groups had set the pattern for medieval culture. The same scholarship is also inconsistent because its implicit praise of the Cathars' hostility to matter is delivered from the pedestal of modern material comfort that science makes possible.

Science may become its own boomerang through modern man's worship if not of matter, at least of material goods. But in order to come into existence, modern science presupposed a positive attitude toward matter, the very attitude that could not be provided by the Cathars, or even by classical antiquity. There mechanical arts were at best tolerated by those in the know, not only the humanist brand but

39. It also prompted the first unambiguous and emphatic statements, in Athanasius's defense of the Logos's divinity, that a universe created by the Father through such a Logos had to be fully logical, that is, rational. For details, see my *Savior of Science*, pp. 75–79.

40. Thus, for instance, Jean Blum, *Les Cathares: Mystères et initiation* (Paris: Le Léopard d'Or, 1985). Anne Brenon, *Le vrai visage du catharisme* (Portet-sur-Garonne: Editions Loubatières, 1988).

also their scientific counterparts, such as Archimedes, to mention only a leading figure.[41]

Mechanical arts were, however, cultivated in a spirit of euphoria through the better part of the Middle Ages. Witness the celebration of progress in the writings produced in the School of Chartres, the speculations about machines that fly, about machines that are driven by men pedaling inside them—speculations in which Friar Roger Bacon was not alone, though certainly best remembered.[42] Those speculations rose to such heights as to inspire designs of parachutes. About 1010, Eilmer, a monk of Malmesbury Abbey, flew with a glider over 600 feet, an event well remembered for almost three hundred years.[43]

Euphoria about material progress found a most colorful expression in medieval cathedrals. Today the plethora of statues and stone decorations in those cathedrals have their exuberance muted by a uniformly grayish shade. Relatively little remains of the original stained-glass windows aflame in inimitable colors. Those windows were matched (the Sainte Chapelle in Paris is the only major surviving example) by the no less colorful painting of columns, arches, and wall spaces.

Nature was fully inside those buildings dedicated to the supernatural. Nature was even celebrated there though never worshiped. A restrained wonder and respect for nature was the only attitude compatible with the doctrine of its having been made by the hands of the Creator. The same doctrine also gave people confidence that as the God-appointed stewards of nature they had to subdue, control, and investigate it. Irresponsible attitudes toward nature have crept in to the extent that people have turned away from the beacon of that doctrine.[44]

41. See Plutarch's "Life of Marcellus," in *Plutarch: The Lives of the Noble Grecians and Romans*, trans. John Dryden and rev. A. H. Clough (New York: Modern Library, n.d.), p. 378. Plutarch also recalls that Plato once reprimanded an astronomer who resorted to compass and ruler to solve a problem in geometry (p. 376).

42. In his *De secretis operibus* (chap. 4) Bacon foretells the construction of ships and wagons that move with incredible speed, as well as of submarines and flying machines.

43. Lynn White, Jr., "Eilmer of Malmesbury, an Eleventh-Century Aviator: A Case Study of Technological Innovation, Its Context and Tradition," *Technology and Culture* 2 (1961), 97–111.

44. A point wholly ignored in the article by Lynn White, Jr., "The Historical Roots of Our Ecologic Crisis," *Science* 155 (1967), 1203–7, which has become a principal cudgel in the hands of some ecologists eager to blame Christianity for the ecological crisis. See my essay "Ecology or Ecologism," a paper delivered at the symposium on ecology and rain forests, organized jointly by the Pontifical Academy of Sciences and the Royal Swedish Academy of Sciences, at the Vatican, May 14–18, 1990, to be published in its *Proceedings*.

Medieval man was, from the natural viewpoint, a motley creature in whose veins ran a wide variety of blood. That this racial mixture produced a particularly creative human being remains forever a vague speculative idea. A quite specific connection seems evident, however, between at least some principal aspects of the technological and scientific inventiveness of those who lived in that age and the faith they commonly shared.

If today mankind is in need of an invention, it is of a nontechnical nature. Late twentieth-century man is in need of reinventing a moral strength whereby he can control scientific inventions that nowadays feed upon one another in an almost runaway process. Moral strength, however, can come only from a morality that is based on knowing the difference between good and evil. Moral strength cannot come from a view of morality within which almost all craving for instant gratification is readily condoned. Furthermore, a morality based on knowing the difference between good and evil also includes knowledge of an all-important point: the knowledge that knowing the good will not necessarily bring about its being done. To forget this is the great fallacy of modern education, increasingly degenerating into mere instruction.

The medievals were no angels, but at least they did not wish to abandon that high level to which their Christian faith had raised their vistas. On that level they held fast to knowing the difference between good and evil, even if all too often they fell far short of doing the good. Nothing is more important for their late twentieth-century counterparts, tied to science hand and foot, than to rescue themselves from the lowlands of moral relativism. They, and especially those in the United States, will not overcome the closing of their minds if a brief recall of places of learning, where an antidote to that relativism is still offered, is all they are willing to take for a remedy.[45] And if they find distasteful the digesting of that remedy, Aristotelian and Scholastic philosophy, they may still take a careful look at the Middle Ages. In Western history that was the first and thus far the last major epoch in which broadly shared respect was paid to the fundamental difference between ends and means.

Means, however spectacular technologically, do not justify the end. Rather they call for an end or a goal that can be morally justified. The survival of our vaunted scientific culture depends on that considera-

45. I have in mind, of course, A. D. Bloom, *The Closing of the American Mind* (New York: Simon & Schuster, 1987).

tion, so unmistakably medieval. If we do not wish to help turn this most scientific age of ours into the most barbaric of all ages, we had better stop using the term "medieval" as synonymous with obscurantist. In doing so, we may make our mental eyes more sensitive to that light which comes from the Middle Ages. It is a light whose spectrum reveals creativity and inventiveness even with respect to science and technology. It does so because as a light it wanted only to participate in the light of true religion. The latter, if true, must have for its primary source the doctrine, nay the very dogma, of the creation of all in the beginning. Aided by that dogma, people of the Middle Ages made the all-important beginning toward science and technology as we know both in our modern times.

Suggestions for Further Reading

Crombie, A. C. *Medieval and Early Modern Science*. Rev. 2d ed. Garden City, N.Y.: Doubleday Anchor, 1959.

Dales, R. C. *The Scientific Achievement of the Middle Ages*. Philadelphia: University of Pennsylvania Press, 1973.

Duhem, Pierre. *The Origins of Statics: The Sources of Physical Theory*, trans. G. F. Leneaux, V. N. Vagliente, G. H. Wagener, with an introduction by S. L. Jaki. Dordrecht: Kluwer Academic Publishers, 1991. The original French was published in 1905–6.

On the Threshold of Exact Science: Selected Writings of Anneliese Maier on Late Medieval Natural Philosophy, ed. and trans. with an introduction by S. D. Sargent. Philadelphia: University of Pennsylvania Press, 1982.

3 Technē and the Origins of Modernity

STANLEY ROSEN

I

In this essay, I make a suggestion about the role of Platonism in the development of the modern epoch. It has become quite common to find in the scholarly literature the claim that such originators of modernity as Copernicus, Kepler, Galileo, and Descartes were influenced by Plato's geometrical cosmology. According to this view, the mythical description of the cosmos as constructed from a variety of pure geometrical figures, which Plato records in the *Timaeus,* is the prototype for the late Renaissance and early modern conviction that philosophy is written in the book of nature in the language of mathematics.

This language, as Galileo puts it, consists of characters that are triangles, circles, and other geometrical figures.[1] Those who, like Ernst Cassirer, Alexandre Koyré, and E. A. Burtt, see in this and similar texts the influence of Platonic metaphysics on the founders of modernity take the passage just cited to refer to the presupposition of a mathematical ontology. The ontological thesis has been chal-

---

1. The quote is from *The Assayer;* the passage is cited in Gary Hatfield, "Metaphysics and the New Science," in *Reappraisals of the Scientific Revolution,* ed. D. Lindberg and R. Westman (Cambridge: Cambridge University Press, 1990), p. 130.

69

lenged by other scholars, such as Stillman Drake and, more recently, Gary Hatfield.

Drake denies that Galileo's science owes anything to the philosophers: "All the documents from his early years at Padua indicate his interest in technological rather than philosophical questions."[2] Hatfield contends that Galileo alludes not to a mathematical ontology but to those properties of natural beings that lend themselves to scientific knowledge: "Mathematics is the language, not of the universe, but of philosophy."[3] Hatfield buttresses his interpretation by citing the *Letter on Sunspots,* in which Galileo tells us to "forsake essences and focus on geometrical properties."[4]

The disagreement between the two sets of scholars suggests that Galileo was either a theoretical Platonist or a practical engineer and scientist for whom philosophy was of interest, if at all, from an epistemological rather than a Platonist standpoint. This disagreement may be generalized to cover most if not all of the founders of modernity.[5] The alternative is theoretical Platonism or an apparently un-Platonic empiricism, with the latter connected to a mathematical epistemology.

My suggestion leaves this controversy unresolved. It is designed instead to broaden the perspective within which we attempt to account for the origins of modernity. The suggestion is: whether or not Galileo and the other founders of modernity were theoretical Platonists, their emphasis on production is entirely compatible with the Platonic dialogues. In a somewhat more elaborate form, I suggest that the difference between Plato and Aristotle with respect to the classification of the arts and sciences has a role to play in the background to the emergence of the modern epoch.

This course leads me to deny, or at least to qualify, the customary opposition, as stated by Alexandre Koyré, that whereas mathematical physics finds its ancestry in Plato, the claim that physics is built directly upon experience and sense perception is Aristotelian.[6] I am, of course, not proposing to deny the link between Aristotle and scientific empiricism. The difference between Plato and Aristotle that in-

2. Stillman Drake, *Galileo at Work* (University of Chicago Press, Chicago: 1981), p. xxi.
3. Hatfield, "Metaphysics and the New Science," p. 130.
4. Ibid., p. 132.
5. Hatfield grants that Kepler was motivated by a theoretical or ontological Platonism; see p. 109 in the cited essay.
6. See Alexandre Koyré, *Metaphysics and Measurement* (Cambridge: Harvard University Press, 1968), pp. 13–15.

terests me here does not concern the role of experience in natural science but rather the relation of production to practice.

II

In the Platonic dialogues, we regularly find classifications of the arts and sciences in terms of whether or not they issue in the production of a separate artifact. Sciences such as arithmetic and geometry produce nothing but discover the natures of mathematical beings; following Plato's terminology, let us call these the theoretical or "acquisitive" sciences. Other forms of knowledge lead to the production of artifacts, as in the case of such arts as carpentry and shoemaking. Plato calls these the "productive" *technai*.

As is obvious from the distinction between theory and production, the Platonic dialogues do not distinguish a third group of practical arts and sciences. Plato does not group ethics and politics as a distinct family that is neither theoretical nor productive. Strictly speaking, there is no ethical virtue in Plato, as is evident from the *Republic*. I here limit myself to the citation of two crucial passages.

In 6 500d4-8, Socrates says that the philosopher "makes" or "fashions" *(plattein)* not only himself but "temperance, justice, and all of demotic virtue" as copies of the Ideas. In 7.518d9ff., he explains that, with one exception, what are called virtues are closer to the body than to the soul and are produced by custom and exercise *(ethesi kai askēsesin)*. The exception is the virtue of *phronēsis*, or intelligence, which is altogether more divine than the others.

The distinction between the genuine or philosophical virtue and the demotic or vulgar virtues is equivalent to the distinction between intelligence and what Aristotle calls the ethical or practical virtues. Plato makes no distinction between theoretical and practical intelligence at the level of philosophy. Instead, he distinguishes between theoretical and productive uses of intelligence. In the light of this distinction, politics is a productive rather than a purely theoretical art.

I say that politics is not "purely" theoretical in order to capture an ambiguity in the Platonic texts. This ambiguity may be illustrated by a reference to the dialogue *Statesman*. The main spokesman of this dialogue, the Eleatic Stranger, engages in a long and extremely tangled discussion, full of obscure divisions and containing an equally obscure myth, the upshot of which is to treat politics as both theoretical and productive. The initial distinction of the sciences and arts is between the practical and the gnostic. The gnostic *technai* are stripped of *praxis;* they produce nothing, neither *pragmata* nor *erga,* that is, neither things nor deeds. The practical *technai* such as carpentry and the handicrafts help to complete bodies that did not previously exist (258d4-e7).

The question now arises: where does Plato classify politics? According to the official diaeresis of the *Statesman,* it lies within the gnostic branch, although the Stranger contradicts himself by placing it first within the purely critical or discriminatory subbranch (260b3) and later in the *epitactic* subbranch, that is, among such arts as architecture, which command or arrange in an order (292b9).

Commanding arts like architecture do not themselves produce but rather give orders for production. Politics is such an art and therefore exemplifies what one may call the intersection between theory and production, but not at all what Aristotle is the first to call practice. This is why the art of weaving plays a central role as a paradigm of the art of the statesman (for example, 305e2ff.). But weaving is a productive, not a theoretical art. We could, of course, say that weaving is instrumental to *phronesis,* which is identified as the royal or political art par excellence (293c5ff.). But we should note that the concluding summary of the dialogue begins with the assertion that we are speaking of "the synthetic sciences" and their *pragmata,* that is, their practical productions (308b10ff.).

These practical productions are, of course, both the city and its citizens. The materials from which the statesman weaves the web of the city are human beings (308d1). It belongs exclusively to the statesman and good lawgiver, acting on behalf of the Muse of the royal art, to produce *(empoiein)* within the souls of correctly blended citizens of diverse natures the true doxa (309d1ff.). And this is also the doctrine of Plato's most famous and comprehensive discussion of the relation between philosophy and politics, namely, the *Republic.*

It would take me too far afield to illustrate with detailed citations the productive nature of philosophical politics in the *Republic.* Having already done this elsewhere,[7] I restrict myself here to the general statement that in the *Republic* Socrates regularly refers to the activity of the philosopher-kings as poetic or demiurgic (for example, 3.394b9-c1: the guardians are restricted to being demiurges of the freedom of

7. See my book *The Quarrel between Philosophy and Poetry* (New York: Routledge, 1988), especially chaps. 1 and 3.

the city). The productive nature of politics also follows from the sickness of the human soul (*Gorgias* 505a6ff.), that is, the rule by nature of desire over spiritedness and intelligence, in the terms of the *Republic*, which must be corrected by the medicinal art of philosophy.[8] The city, in other words, particularly the healthy city, does not exist by nature. It must be produced by a theoretico-productive art whose paradigms include weaving and medicine.

The true statesman does not produce directly, but he does so indirectly, either by formulating laws and customs (as in the *Republic*) or by ruling over those who are capable of producing a city of virtuous citizens (as in the *Statesman*). The reason why it is necessary to produce the city is because man is naturally defective, and even sick; this defect or illness can be cured, or else contained, by the art of weaving together human beings whose natural differences correspond to the warp and woof of a cloak, blanket, or other woven artifact.

In the *Statesman,* theoretical production is not addressed to the mastery of the cosmos but rather to the therapy or care *(therapeia, epimeleia)* of human beings. This care is illustrated by the paradigm of weaving, particularly the weaving of wool into various shelters or garments for protecting the body against the rigors of nature (279c7). In sum: politics is a theoretico-productive art that corrects nature or supplements it with artifacts for the protection of human beings. Politics is both productive and defensive. I can now express the fundamental difference between the ancients and the moderns in political terms. The ancients construct tools as a defense against nature; the moderns subscribe to the thesis that the best defense is a powerful offense.

By putting the point in this way, I do not at all intend to overlook or minimize the importance of radical new discoveries in mathematics, physics, engineering, and the other natural sciences. But even if these discoveries were all made independently of more general philosophical reflection, which I doubt, it remains true that the coordinated plan to employ them for the safety and increased comfort of the human race, and still more fundamentally to make humankind the master and possessor of nature, is political in the comprehensive sense of that term.

In this light, one may suggest that the incomplete separation of theory from production in Plato, as well as the absence of a distinct

8. Cf. the discussion of ugliness and disease in *Sophist* 228a4ff. with the comparison between philosophy and medicine in *Gorgias* 464d2ff. and 500e4ff.

buffer zone of the practical, makes it easy to conceive of the fundamental paradigm of the modern revolution: a theoretically articulated
project, initiated by the human will, to transform human life through
the instrumental mastery of nature. This is much closer to the Platonic paradigm of philosophy as the unity of theory and practico-
production, which culminates in the prophetic or lawgiving activity of
the philosopher, than it is to the Aristotelian paradigm of theoretical
philosophy as independent from practice, and of both as independent
from production.

The modern epoch does not originate in a shift from Aristotelian
theory to a production that is equally Aristotelian in its distinctness
from practice. It would be more accurate to say that modernity origi-
nates in a shift from the Aristotelian tripartition of the arts and sciences to a revision of the Platonic model in which, despite the
fundamental distinction between knowing and making, the lines
separating theory, action, and production are tangled and blurred.
The tangled state of these lines thus constitutes a genuinely Platonic
presupposition for the eventual assimilation of theory into practico-
production in the modern epoch.

Let me emphasize that I am not attempting the anachronistic task
of making Plato into a modern, nor have I forgotten the differences
between Platonic *theoria* and modern constructive theory. My suggestion is intended to broaden the terms within which discussion of the
origin of modernity takes place. Hitherto it has been popular, even
in the face of recent criticism, to present the origin of modernity as
a renaissance of Plato's mathematical *theoria*. I leave this claim as it
stands, suggesting instead that the deeper influence of Platonism on
early modernity is practico-productive.

Allow me one more reference to the *Republic* in an effort to clarify
and sharpen this suggestion. In Book Ten, Socrates observes that
there are three *technai* that pertain to all other arts: using, making,
and imitating. Of these three, it is the *technē* of the user that determines the excellence, beauty, and rightness of everything that is made
or that exists by nature (601d1ff.). The user is, of course, man, and
the useful is relative to human intentions, which are independent of
the mathematical or eidetic structures of things even if conditioned
by them.

The subordination of making and imitating to usefulness is the
core of the union between production and practice. It is no doubt

true that in Plato human intentions are subordinated to the natures of things, and also to human nature. But this in itself does not constitute a difference between Platonism and modernity. It will take several centuries for the subordination of form and essence to mathematics, and of mathematics to the human will, to dissolve the Platonist assumption of the founders of modernity that science discovers rather than makes the natural order.

One of the crucial steps in this process, in my opinion, is a shift in the understanding of happiness. Here again I believe that I see the subterranean influence of Platonism at work. According to Aristotle, happiness is possible not only for the theoretical man but also for the practical man, albeit in two distinct senses. In Plato the possibility of happiness is considerably more obscure. Since there is no ethical or practical virtue apart from theoretical virtue, and if happiness depends upon perfection, that is, upon the possession of virtue, then the happiness of the nonphilosopher must be specious at best. One might object that it is guaranteed by the rule of the philosopher, but this rule will occur, if ever, only in the beautiful city of speech that is constructed in the *Republic,* and the possibility of this city is, to say the least, dubious.

As to the philosopher himself, there is good reason to suspect that he also cannot be truly happy in the imperfect city; to the extent that such happiness is available, it seems to be associated with a radical detachment from political or communal existence, and so with assimilation into a pure intellectual contemplation of eternal forms. On balance, one must say that the Platonic dialogues lean toward the teaching that human life is unhappy. The sickness and restlessness of the desires are amenable only to a perception of the good or, as one may also express this, to a perception of the goodness of reason.

When reason becomes separated from the good, as happens in early modernity under the influence of an un-Platonic reliance upon the exclusive paradigm of mathematics, the desires are no longer subject to rational restriction. Geometrical structure and algebraic ratio carry no sense of temperance or of the nobility of reason. There is no longer any reason to restrict the desires, not even the desire to conquer, and thus change, geometrical and algebraic order. Happiness is then reinterpreted as the satisfaction of desire, and satisfaction is shifted into the future. As Leo Strauss once said of John Locke's doctrine of man, life is the joyless quest for joy. To which I would add: how dangerously close to Plato!

III

So much for my general sketch of the suggestion that the practico-productive dimension of Platonism has an important role to play in the origination of modernity. I now turn to a corollary thesis, namely, that the modern view of the unending perfectibility of *technē* was accessible to the ancients as well. We have just seen that in the Platonic dialogues, *technē*, whether productive or imitative, is subordinated to human intentions on the one hand and to *phronēsis*, or sound judgment on the other. This is quite different from the claim that *technē* is subordinated to the Ideas, and, in fact, I know of no occurrence of that claim in the Platonic dialogues.

The Ideas serve as the *telos* of philosophical Eros or, in an alternative formulation, of pure theoretical cognition. They may be said to determine what is truly, and so what can be known, but they do not directly determine what we ought to do or make. This determination depends entirely upon the mediation of the philosopher-king, and thus upon theoretical rather than practico-productive excellence.

Even here, it is not so much the direct apprehension of pure forms that guides the political activity of the philosopher as it is his intellectual judgment concerning how human beings ought to live if those who cannot see the Ideas are to be preserved in an orderly and temperate community. As to the excellence of such a community, it is determined in the final analysis by the degree to which the philosophical life can take place within it. As we have already noted, the nonphilosophical or demotic justice is associated with the body, not the soul. In the case of the soul, justice is equivalent to *phronēsis*, or intelligence, the same term that occurs in the *Statesman* to designate the perfect form of the royal art.

All this being so, it follows that the intentions of the philosopher, who is for Plato the highest type of human being, ought to direct all of the arts and sciences. This direction should be in the light of the Ideas, or what Socrates calls the light of the highest Idea or principle of the Ideas, the Good (518d9ff.). The Good is not itself an Idea in the sense of a form of this or that kind of being; Socrates compares it to the sun that illuminates all things on the face of the earth and thereby causes them to grow as well as to be visible.

It would not be easy to show that the Good serves a political or ethical role in addition to what we may call its ontological function as the principle of beingness, or *ousia*. Stated as sharply as possible, the Good does not tell us what to do. What it does is to illuminate the Ideas toward which we, that is to say, in the first instance the philosophers among us, are directed by Eros. From all points of view, I believe, we arrive at the conclusion that human activity, on Plato's analysis, ought to be regulated by the philosopher, who is in turn guided in his intentions by the love of wisdom. The philosopher is guided in his intentions by philosophy.

To return to the case of *technē*, it is not difficult to understand the sense in which it is to be regulated by philosophy. In the *Republic*, for example, *technē* is regulated in order to preserve the rule of the philosophers, which in turn is required if the city is to be just. But why do we require a just city? In order to be happy, which in turn depends upon a well-regulated soul. And this can be acquired only if philosophers are kings and kings philosophers.

Strictly speaking, it is false to say that justice, except as a synonym for philosophy, is for Plato a good in itself. The most that could be claimed is that happiness or blessedness is a good in itself, and that this requires us to be just. This amounts to saying that philosophy and happiness are synonymous, or, more narrowly, that they are synonymous in the city that allows the perfect philosopher to exist.

In actual or historical cities, the reasons for restricting the progress of *technē* are not so compelling. I do not mean by this that Plato was in favor of unlimited technical progress as a principle of his ordinary political doctrine. No doubt he was quite conservative on this point, as on many others. But if the philosopher does not rule, then the city cannot, as it were, be abstracted from history. In all actual cities the happiness available to the philosopher must be entirely private, as I have noted previously.

This is not contradicted by the extraordinary emphasis on politics in the Platonic dialogues. That emphasis is due in the last analysis to the conflict between philosophy and politics, or to the inexpugnable conflict between philosophical and nonphilosophical conceptions of happiness. It is this conflict that leads us to doubt whether happiness is available to human beings of any kind, whether philosophical or not, given Plato's understanding of our natures.

One might therefore listen to the conversations of Socrates, or somewhat later read the Platonic dialogues, and come away with the feeling that philosophy is irrelevant to human well-being, let alone to human happiness. And this doubt places the question of *technē*, in quite a different light. Or rather, it shifts us back to the light of ordinary political experience, as that light is cast upon the pages of Thucydides' *History of the Peloponnesian War.*

Thucydides is silent about philosophy in his great inquiry; the word appears only once, in the speech of Pericles, where it does not carry the Platonic sense but means something like *paideia*, or musical and verbal cultivation. If the fundamental fact about human life is war, then it is impossible to restrict the development of *technē*. This problem is not discussed by Socrates in the *Republic*; despite the crucial role of the guardians, we are given the impression that war may be limited if not entirely avoided, together with what is today called "international relations."

Thucydides presents us with the actual political situation in book 1, chapter 71, section 3. The context is the debate between the Corinthian and Athenian representatives before the Spartan authorities in the preliminary stages of the outbreak of the war. The Corinthians contrast the daring and innovative nature of the Athenians with the slow, cautious, excessively traditional nature of the Lacedaimonians, and warn that the necessities of political change require technical innovation. It is necessary in politics "just as in *technē* always to master what comes next."

Should one not then say that it is Thucydides rather than Plato who provides us with the classical antecedent of the modern unleashing of *technē*? Perhaps so, but this is a corollary to my original thesis, not its contradiction. My suggestion was not that Plato unleashes *technē*, but that he does not, in fact, distinguish as sharply as he seems to do between the theoretical and the practico-productive *technai*. As soon as Thucydidean realism triumphs over Platonic madness, Plato's own tangling of the lines separating politics from theory on the one side and production on the other serves as a philosophical basis for the articulation of realism into a comprehensive program in which politics and philosophy are married by experimental and mathematical science.

As I see it, in the passage cited above Thucydides expresses what is implied or implicitly granted by Plato's attempt to subordinate politics to philosophy, an attempt that is, for Thucydides, impossible and not even worth considering. He silently rules out the possibility of the regulation of desire by intelligence. This being so, and regardless of whether his personal political preferences are conservative or progressive, the subordination of *technē* to political control must be rejected as impolitic. Add to this the Platonic notion of politics as a productive art, and at least part of the stage is set for the advent of modernity.

Given what I call the Thucydidean inflection of Platonism, the intentions of the philosopher are directed away from imaginary principalities toward the actual task of defending humankind in its war with nature. In other words, nature is war, not peace. In ontological terms, the mathematical structure of nature is subject to the will of humankind in its pursuit of comfortable security. Ethics is replaced by physiology, and happiness gives way to the unending effort to satisfy desire.

The union of the view that life is fundamentally war with the Platonic understanding of politics as a productive *technē* gives birth to the decisively modern conception that human beings are, or may become, the masters of their destiny. As a crucial text to illustrate this conception, I cite a well-known passage from Machiavelli's *Prince:* "It is better to be impetuous than respectful, because fortune is a woman, and it is necessary, if you wish to keep her quiet, to beat her and hit her. It is evident that she lets herself rather be conquered by such men than by those who proceed coldly. Therefore, like a woman, she is always the friend of the young, because they are less cautious, more ferocious, and command her with more audacity."[9]

Machiavelli is speaking here of politics, not of natural science. But my point is precisely that there is a political presupposition or context for the extension of the productive model of *technē* that is found within the natural sciences. Perhaps it would be more accurate to say that the political and scientific senses of technical production modify and stimulate each other so as to lead to the development of a comprehensive philosophical program of revolution, a revolution that can no longer be designated as either theoretical, practical, or productive because it is all three in one.

This passage from Machiavelli should be read together with Descartes's *Passions of the Soul,* in which, as one could say, ethics and politics are united with natural science through the medium of physi-

9. *Il principe,* in *Il principe e Discorsi* (Milan: Feltrinelli, 1960), p. 101.

ology. Descartes instructs men of strong or great souls (what the ancients called *megalopsychia*, which he reinterprets as *générosité*) to destroy the belief in fortune by reflecting upon divine providence as equivalent to the natural order.[10] This order is both necessary and accessible to human intelligence. It cannot itself be changed but may be understood and so mastered in the sense that it may serve as the basis for supplementary constructions that will safeguard and improve the quality of human existence.

There is then no "mathematical ontology" in Descartes; mathematics is the language constructed by humankind to further the intentions of the will of the philosopher. It would also be a distortion of Plato's thought to infer from the geometrical cosmology of the *Timaeus* that he held to a mathematical ontology. At the same time, I know of no compelling evidence in the Platonic dialogues to sustain the thesis that the Platonic Ideas are arranged in a hierarchy of excellence, or what Nietzsche called a *Rangordnung*. There is no obvious sense in which the various accounts of the Platonic Ideas are connected with a doctrine of teleology. On the contrary, the order of excellence in the cosmos is usually attributed by Plato's interlocutors to the *epimeleia* of the ranked Olympian gods, led by Zeus, who order and care for the cosmos.[11]

This attribution means that the intentions of the philosopher, despite his Eros for the Ideas, are not determined directly by them, but by the philosopher's *phronēsis*, or judgment, as to what action best serves the interests of philosophy. Such a view comes much closer than the Aristotelian teleology to the modern doctrine of the freedom of thought from regulation by natural ends, even while respecting the natural order. One has only to replace the Platonic Ideas with the geometrical structures of the *Timaeus* or, more radically, with the symbolical beings of modern mathematics, and the philosopher is transformed into a mathematical physicist free to do whatever he believes will satisfy his desire for the truth. And this is a step away from the desire to satisfy desire.

That Aristotle is not so close to modernity on this point may be suggested by the following citation from the *Politics*. In Book 1, at 1257b25ff., Aristotle says: "The art of medicine pursues health without limit, and each of the arts pursues its end in an unlimited way

---

10. *Les passions de l'âme*, ed. G. Rodis-Lewis (Paris: J. Vrin, 1955), p. 170.
11. See, for example, *Phaedrus* 246d6ff., as well as *Statesman* 271c3ff.

(for they wish to accomplish this to the greatest possible extent); although with respect to the end they do not proceed in an unlimited way (for the end is a limit to all the arts)."

In other words, *technē* has no intrinsic limit with respect to the development and perfection of means to the accomplishment of its end. This is especially plain in the case of war, where each stage in the development of weaponry forces the opponent to match and surpass it. But according to Aristotle, there is an external limit to the progress of *technē*, namely, the end toward which it is directed. Peace is the natural limit to war, and health is the natural limit to medicine. What does this argument amount to? It seems to license unending technical perfection, with the nominal qualification that this perfection is always in the service of an unchanging end or limit.

This is a fragile argument for more than one reason. If technical progress is allowed to continue endlessly, a time will come, as we have reason to know, when the ends will themselves be dissolved by an unending process of interpretation. In addition, progress in the development of weaponry may come to a temporary end with the conclusion of one war, yet it starts up again with the outbreak of new hostilities, to say nothing of secret research during periods of ostensible peace. And an analogous argument may be constructed in the case of medicine. For all practical purposes, this means that the ends do not serve to limit technical progress but at the most to identify it as progress of such-and-such a type. As we also have reason to know, these types are not so easy to distinguish; the line between medical and military research cannot be drawn cleanly on the basis of a view toward the telos of health.

For a text from the Aristotelian school, if not directly from the master himself, that is closer in keeping with the origins of modernity, we may cite the opening lines of the treatise called *Mechanical Problems* (*Mēchanika*), usually attributed to the post-Aristotelian peripatetic school. This text begins by distinguishing between "marvelous events that occur in accordance with nature, of which we do not know the cause, and those that occur contrary to nature, which come to be through *technē* for the benefit of mortals. For in many cases nature makes something that is contrary to our benefit. For nature always acts in the same way and simply; but what is beneficial changes frequently."

This is extraordinarily modern in spirit. The author goes on to say that we must have recourse to *technē* in order to produce a "device" *(mēchanē)* that will enable us to act contrary to nature. "For as the poet Antiphon wrote, and as is true, 'we master by *technē* those things in which we are conquered by nature'" (847a11–21).[12] It reminds us of the Cartesian radicalization of Bacon's New Instauration, or the turn to a practico-productive theory with the power to make men the masters and possessors of nature.[13] I have been suggesting that the middle term between Antiphon and Descartes is Plato rather than Aristotle.

IV

According to Sir Francis Bacon, the greatest obstacle to progress in the sciences lies "in the despair of mankind and in the supposition of its impossibility."[14] A parallel remark is to be found in Descartes's *Passions of the Soul.* In speaking of desires that depend upon ourselves for fulfillment, Descartes says: "The fault that one is accustomed to commit here, is never to desire too much; it is only that one desires too little."[15] The possibility of scientific progress is implicit in *technē*, that is, in the proceedings of the artisans, which we must employ to discover the force and the actions of the elements and the heavens.[16] This is an example of activity that depends upon ourselves.

What one could call Platonic pessimism, or a despair of the possibility of happiness, lightly disguised by the salutary equanimity of a Socrates "grown beautiful and young,"[17] is transformed into the optimism of the founding fathers of modernity by a series of steps, some of which, paradoxical though it may sound at first hearing, have been furnished by Plato himself.

Unlike the analogous situation in Aristotle, the cosmos of genesis is, according to Plato, not eternal but transient and cyclic. This has to do with the famous and obscure "separation" of the Platonic forms from their generated copies or instances, as well as with the reduction of virtue to wisdom and the correlative assimilation of practico-production into the domain of philosophical *phronēsis.*

12. I am indebted to David Lachterman for calling this passage to my attention.

13. *Discours de la Méthode,* ed. E. Gilson (Paris: J. Vrin, 1947), p. 61f.

14. *Instauratio Magna,* in *Works,* ed. Spedding, Ellis, and Heath (Boston: Brown, 1861), 1:415. See p. 406, where Bacon says that human beings tend to underrate their strength.

15. *Les passions de l'âme,* p. 177.

16. *Discours,* p. 62.

17. *Epistles* 2.314c4.

None of this is intended to suggest that modernity is a direct consequence of, or even that it is somehow at bottom fully compatible with, Platonism. Perhaps I could summarize this essay by saying that I sense in Plato a poetical discontent with the place of human beings in the cosmos that I do not not find in Aristotle. From this standpoint, I would not regard the aporetic nature of so many of the Socratic conversations, or the reliance upon myth in the decisive cases of the soul, the city, and the cosmos, as concealing a deeper positive teaching but rather as presenting a lightly masked representation, couched in the tones of Attic urbanity, of the incipient spirit of modernity.

I close with a prudent reminder: as these terms are normally employed, Plato was an ancient, not a modern. I mean by this something more than the obvious chronological or historical remark. The practico-productive nature of Plato's political philosophy is restrained by the Ideas on the one hand and the radical difference between the philosopher and the nonphilosopher on the other. There is no point in speculating upon what Plato would have said, had he been aware of modern mathematics and natural science. Yet I cannot entirely restrain the impression that he would have aligned himself with the founders of modernity, less because of the spirit of philanthropy than because of his deeply poetical soul.

To say this in one more way, the "dialectic" of the dialogues, understood as the aporetic effort to grasp the visible yet illusive Ideas, and as distinct from the (in my opinion) imaginary dialectic of pure Ideas that Socrates wishes for in the *Republic*, is closer to what Hegel called "the peculiar restlessness and dispersion of our modern consciousness"[18] than is the Aristotelian effort to imitate the "thought thinking itself" of the god of the philosophers. I say this despite Hegel's own marked preference for Aristotle, upon whom he clearly modeled himself, and also despite what is to my mind the finally unconvincing portrait in Plato of philosophical *Heiterkeit*, or "beyondness," a portrait that is contradicted by the actual description of human existence, including the destiny of the philosopher, for whom life is a preparation for dying.

Whether this impression is right or wrong, however, has no bearing upon the soundness of the suggestion that I have outlined in this

18. *Wissenschaft der Logik* (Leipzig: Felix Meiner Verlag, 1951), 1.20.

essay. Plato's contribution to the origin of modernity is closer to *technē* or *poiēsis* than it is to mathematics.

Suggestions for Further Reading

Blumenberg, Hans. *Die Legitimität der Neuzeit.* Frankfurt: Suhrkamp Verlag, 1966.
Lachterman, David R. *The Ethics of Geometry: A Genealogy of Modernity.* New York: Routledge, 1989.
Rosen, Stanley. *The Ancients and the Moderns: Rethinking Modernity.* New Haven: Yale University Press, 1989.
———. *The Limits of Analysis.* New Haven: Yale University Press, 1980.

4 Science and Freedom: America as the Technological Republic

WILSON CAREY MCWILLIAMS

From the beginning, Americans, in appreciation of technology's comforts and conveniences, its relief from drudgery and want, have honored the nation's great inventors. This is not surprising: even technology's critics are appalled by hunger's shadow on a child's face, and a plague such as AIDS turns laboratories into temples of hope. But technology also brings nuisances and deadly perils, and that dark side is something Americans have always known or suspected. Our schools teach that the cotton gin strengthened slavery, with the implicit moral that the machine can corrupt the garden as subtly as the serpent.[1]

Technology is more than gadgets and machines: it is associated with a frame of mind, a preoccupation with ways of getting results, a dangerous spirit that strains against limits and for rule. And America's oldest wisdom teaches that when the people of Babel set out to build a tower to heaven, the Lord scattered them and confounded their speech, political disaster following technological pride.[2]

From their religion, early Americans learned that human beings

1. Leo Marx, *The Machine in the Garden* (New York: Oxford University Press, 1964).
2. Genesis 11:4–9.

are naturally subject to limits, parts of a good whole. Men and women are granted "dominion" only within Creation and subject to rule; the alienation of humanity from nature, like the estrangement of human from human, is the result and mark of sin. According to this doctrine, the goal of human life and prudence is to achieve, as far as possible, the reconciliation between human beings and nature and, finally, to reconcile humanity and God. Redemption is ultimately the work of grace, but law, religion, and family, all the great institutions of human education, are naturally intended to promote a loving attachment between human beings and their fellows and an adoring recognition of the wonder of the created order.

Science and technology, consequently, were to be cherished insofar as they revealed the excellence of God's design, and to that extent, there was an affinity between early American religion and the new science. Even champions of orthodoxy supported technologies, such as vaccination, that bettered human life, provided that the highest honor was reserved not for the device but for God as first cause.[3] Like all human works, science and technology were to be ruled by right subordination and attachment, governed by nature and the good life.

The strongest voices in early American religion rejected asceticism. Calvin argued that "delight," God-created, was meant to aid the soul, and John Wise maintained that God intended human beings to "banish sordidness, and live Bright and Civil, with fine Accomplishments."[4] But religious teaching also held that there are limits to what is needful, however difficult it is to define them in practice; beyond that boundary, wealth is luxury, a distraction to the soul. Most important, the doctrine of sin testified emphatically that some powers are not safe in human hands. The first principle of politics is the education and right government of the soul, so that implicitly the excellent soul sets the rule that measures the arts and the laws.

Human beings are made happy not by having much but by needing little, and Protestant teaching sought to develop the inner qualities necessary to make the best of any station, including the finite, human place within nature.[5] In Daniel Defoe's great celebration of resourcefulness, Robinson Crusoe saves what he can from his wrecked ship, but this remnant of technology leaves him vulnerable to loneliness

3. Perry Miller, *The New England Mind* (Boston: Beacon, 1961), 2:345–63.

4. John Calvin, *Institutes of the Christian Religion,* bk. 3, chap. 10, sec. 23; John Wise, *A Word of Comfort to a Melancholy Country* (1721), cited in Miller, *New England Mind,* 2:329.

5. Calvin, *Institutes of the Christian Religion,* bk. 3, chap. 10, sec. 6.

and despair. The foundation of his industriousness is his conviction that Providence is pervasive, so that if he bestirs himself, he will find all that is needed for an ordered life. Thus assured, Crusoe successfully champions right order against violations of natural duty (cannibalism) and civil obligation (mutiny), bringing his island a kind of decent governance.[6]

The new science, however, proved less governable, linked as it was to the modern political philosophy that had helped nurture it. In the education of the founding generation, the new sciences of nature and politics were so intertwined that, for Thomas Jefferson, Newton, Bacon, and Locke—the "three greatest men the world had produced"—were a secular trinity, almost as inseparable as the three persons of its Christian rival, and certainly grander than ancient philosophy.[7]

The new doctrine described a natural order, but one lacking moral authority in human affairs. In the new view, nature is matter in motion, its order defined by the mechanics of attraction and repulsion, at best indifferent to human hopes.[8] Human beings are moved by need and desire and by the effort to avoid pain and death; self-preservation, "the first and strongest desire God planted in men," is "the first principle of our nature."[9] Since the human conflict with nature, in these terms, is itself natural, human beings properly seek mastery over nature, a goal that requires them to become supreme mechanics with their hands on the levers of natural power. Technology, in this sense, is the answer to the human problem, worthy to be pursued through all methods and inventions that "let Light into the Nature of Things, tend to increase the Power of Man over Matter, and multiply the conveniences or Pleasures of Life."[10]

6. Daniel Defoe, *Robinson Crusoe* (Philadelphia: J. C. Winston, 1925), p. 167; see also pp. 118–19.

7. Letter to William Duane, August 12, 1810, in *The Life and Selected Writings of Thomas Jefferson*, ed. Adrienne Koch and William Peden (New York: Modern Library, 1944), p. 609.

8. Thomas Hobbes, *Leviathan*, ed. C. B. Macpherson (Harmondsworth: Penguin, 1968), chap. 6, pp. 118–19, and chap. 11, 160–61. "On the basis of sensation, of matter and motion," Jefferson wrote, "we may erect . . . all . . . certainties," rejecting any knowledge except what comes to the body through the senses. *The Adams-Jefferson Letters*, ed. Lester J. Cappon (Chapel Hill: University of North Carolina Press, 1959), pp. 568–69.

9. John Locke, *Two Treatises of Government*, I, para. 86, 87; II, para. 149; Alexander Hamilton, "A Full Vindication" (1774), in *The Works of Alexander Hamilton*, ed. H. C. Lodge (New York: Putnam, 1903), 1:12; see also *The Writings of James Madison*, ed. Gaillard Hunt (New York: Putnam, 1900–1910), 4:387.

10. Benjamin Franklin, *A Proposal for Promoting Useful Knowledge among the British Plantations in America* (1743), in *Benjamin Franklin: Writings*, ed. J. A. Lemay (New York: Library of America, 1987), p. 296.

Nature furnishes the human goal—its own subjugation—but not the moral basis of human society. Benjamin Franklin's early *Treatise on Liberty and Necessity, Pleasure and Pain* (1725) located human beings wholly within the mechanisms of nature, reasoning that, given God's power, knowledge, and benevolence, the human world must be presumed to have the regularity of the natural world. "Man is a part of this great Machine, the Univers," for the "ingenious Artificer" would not have manufactured a second system "endu'd with an independent *self-motion*, but ignorant of the general interest of the Clock."[11] The young Franklin concluded that evil cannot exist; when human beings pursue their self-interested natural inclinations, it must be presumed that they forward the good of the whole.[12] Experience led Franklin to regret having printed this pamphlet because of what he regarded as its bad moral consequences. The fault, he concluded, was not in his logic but in his premises: neither God's nature nor His design is known sufficiently to furnish a basis for moral deduction. The moral order, then, must be seen as a thing of human making, to be judged in terms of its perceptible effects and utilities.[13]

Yet human reason is also an inadequate basis for moral and civil order. As Locke had argued, the fact that reason separates human beings from the rest of nature—making them conscious of the struggle with nature's limits—allows them to fall below the level of instinct as well as enabling them to rise above it.[14] Reason may moderate passion, but at bottom it serves desire. Franklin reports that, despite his vegetarian principles, he persuaded himself that it was not "murder" to eat fish, since fish ate other fish. Yet he went on to indicate, tongue-in-cheek, the instrumental quality of his own reasoning: "So convenient a thing it is to be a *reasonable Creature* since it enables one to find or make a Reason for everything one has a mind to do."[15] Reason can administer, but it cannot be trusted to rule, especially since human passion reaches beyond eating fishes.

In the characteristic political science of the founding era, the state of nature, when human beings have no stay or support but instinct and reason, is open to imperious desires and exposed to desperate

11. Ibid., p. 61; compare Locke, *Essay concerning Human Understanding*, bk. 4, chap. 3, para. 18.

12. *Benjamin Franklin: Writings*, pp. 61, 63–64.

13. Letter to Thomas Hopkinson, July 1748, in ibid., p. 435; Benjamin Franklin, *Autobiography and Other Writings*, ed. Kenneth Silverman (New York: Penguin, 1986), p. 63.

14. Locke, *Two Treatises of Government*, I, para. 58.

15. Franklin, *Autobiography*, p. 39.

dangers, so that—whether in the beginning or soon after—it comes down to a state of war.[16] Seeking to escape that condition, human beings, following the path sketched by social contract theory, agree to create political society, giving up certain rights in order to enjoy the rights they retain more safely and effectively.[17] Nature drives human beings into political society, but politics is not natural; it is a technological project, an instrument for protecting rights and effectuating essentially private desires, properly rooted in the consent of the intended users. Government's lodestar is not virtue but interest.

All human beings are self-interested, in the most general sense of the term, so that for the Framers the public good is equivalent to a "judicious estimate of our true interests."[18] Most human beings, however, define themselves narrowly, thinking of their bodies and desires or, at best, identifying with their families and properties. Human judgment is chronically shortsighted and parochial: "momentary passions and immediate interests have a more active control over human conduct than general or remote considerations of policy, utility or justice."[19] This is especially true because long-term calculations are uncertain and exposed to unexpected events. The "mild and salutary" constraint of law and the assurance of government policy are necessary to prod or prompt us toward our interests, "rightly understood."[20] In the conduct of government, and even more in its founding, the majority of human beings need the guidance of that political minority who identify themselves with public things—power, fame, or the good of their kind—and are thus "disinterested" in the common sense of the word.[21] But these more public motives and goals

16. Hobbes's view of the state of nature, Franklin wrote James Logan (1737?), is "nearer the truth" than the idea that the original condition was a "State of Love" (*Benjamin Franklin: Writings*, p. 425).

17. *Federalist* No. 2, p. 51; *The Papers of James Madison*, ed. W. T. Hutchinson et al. (Chicago: University of Chicago Press, 1962–), 5:83, 8:78, 300.

18. *Federalist* No. 1; or, in Madison's phrase, "the permanent and aggregate interests of the community" (*Federalist* No. 10). When our duties and interests "seem to be at variance," Jefferson wrote to Say (February 1, 1804), "we ought to suspect some fallacy in our reasonings" (*Life and Selected Writings*, p. 575).

19. *Federalist* No. 6; see also Franklin, *Autobiography*, pp. 103–4; even the best measures of government, Franklin remarked, are ordinarily "forc'd by the Occasion" (p. 147).

20. *Federalist* No. 20; see also No. 15. Tocqueville's discussion of interest "rightly understood" may be found in *Democracy in America*, vol. 2, bk. 2, chap. 8.

21. David Epstein, *The Political Theory of "The Federalist"* (Chicago: University of Chicago Press, 1984), p. 6; Douglas Adair, *Fame and the Founding Fathers* (New York: Norton, 1974), pp. 3–25; on the ideal of disinterestedness and challenges to it, see Gordon Wood, *The Creation of the American Republic, 1776–1787* (Chapel Hill: University of North Carolina Press, 1969), pp. 471–518.

have no superior moral title, since at bottom they come down to self-interest, though more amply defined. In fact, such ambitions are dangerous because the desire for power, the quest for greatness, or the zeal to do good are expressions of the yearning for dominion.

Governments, consequently, must be so contrived that public officials represent and administer only private ends and are held accountable by private citizens. As Jefferson wrote in *A Summary View of the Rights of British America* (1774), the king "is no more than the chief officer of the people, appointed by the laws and circumscribed with definite powers to assist in working the great machine of government erected for their use, and consequently subject to their superintendence."[22] The Framers' modern political science rejects *rule*, a government that measures and is measured according to a scale of excellences or virtues, in favor of *rules*, officials qualified by technical skill and "kept within their due bounds."[23]

The right design of the laws is the more urgent because government must be a house of power, forwarding the human effort to master nature.[24] Government, the *Federalist* insists, must possess "energy," power decked in the language of science. It needs force sufficient to overcome obstacles to its authority, to protect civil peace and the rights of citizens, to forestall foreign danger, and in general to protect the sphere of order against the disorder of nature.[25] More positively, government must help to promote the accumulation of power, wealth, and technique, supporting commerce and encouraging "science and useful arts."[26]

Power over nature, however, entails the growth of human power within political societies and between them, with a corresponding increase in the possibilities of abuse. The "science of finance, which is the offspring of modern times," Alexander Hamilton remarked, gives government new control over money, and hence over politics, notably by allowing (and requiring) professional armies, with their

22. *Political Writings of Thomas Jefferson*, ed. Edward Dumbauld (New York: Liberal Arts, 1955), p. 16.
23. Locke, *Two Treatises of Government*, II, para. 137; Sheldon S. Wolin, *Politics and Vision* (Boston: Little Brown, 1960), pp. 347–48; Norman Jacobson, "Political Science and Political Education," *American Political Science Review* 57 (1963), 561–69.
24. I take the term "powerhouse" from *The Education of Henry Adams* (Boston: Houghton Mifflin, 1961), p. 421.
25. *Federalist* Nos. 15, 16, 23, and 70.
26. Article I, sec. 8. *Federalist* No. 12 refers to "multiplying the means of gratification" as a public good.

potential for repression.[27] Similarly, commerce, though generally benign, also has prompted the trade in slaves, that "barbarism of modern policy."[28] Consequently, government needs power in relation to these challenges and outrages; its "means ought to be proportioned to the end" as causes to effects, for "wherever the end is required, the means are authorized."[29] The laws can rule only if they shape necessity or obey it, and either alternative demands power.[30]

Yet although government and order are the preconditions of any effective human freedom, liberty is first by nature, the end and purpose of political life, so that, James Madison argued, safeguarding "the diversity in the faculties of men" is "the first object of government."[31] Since reason is naturally connected to self-love, it is impossible to deal with the "causes" of faction without repression, but that solution is *morally* unacceptable. If faction is the price of freedom, it is one that a rightly constructed government must pay. Factions must be controlled through their "effects," just as human beings must be governed through their behavior, leaving them at liberty but shaping their choices, channeling human desires and actions through institutions scientifically designed to get results.

In fact, when the American Framers appeal to the "improved" science of politics, they are thinking less of theoretical principles than of the very practical technology of institutions, such as representation and the separation of powers.[32] Of course, the Framers recognized that political technology must be adapted to the people and the times to which it is being applied, especially since at the outset, new laws can only point toward a new political culture.[33] And the United States was not a perfect laboratory: the Framers were forced to accommodate to slavery, a violation of natural right, and to certain attachments, such as state loyalty and trial by jury in civil cases, which

27. *Federalist* No. 8.
28. Ibid., No. 42.
29. Ibid., Nos. 31, 42.
30. Ibid., Nos. 25, 41.
31. Ibid., No. 10; Epstein, *Political Theory of "The Federalist,"* pp. 144, 163.
32. *Federalist* No. 9; representation, Madison argued, is a "great mechanical power," discovered by Europeans but one that Americans have applied and can develop to its "full efficacy" (No. 14).
33. Michael Kammen refers to the Framers' statecraft as "a blend of cultural relativism combined with theoretical universalism" (*A Machine That Would Go of Itself* [New York: Knoph, 1986], p. 65). In 1814 Jefferson wrote to Thomas Law that "men living in different countries . . . may have different utilities; the same act, therefore, may be useful, and consequently virtuous in one country which is injurious and vicious in another" (*Life and Selected Writings,* pp. 639–40).

many would have preferred to do without.[34] Still, foreign threats were not pressing, society was prosperous and without extremes of wealth and poverty, and the public was accustomed to personal and political liberty. So favored, America was relatively free to test the new science and to establish whether, in principle, it is ever possible for a people to give itself laws on the basis of reason and choice.[35]

Moreover, American experience and reflection convinced Madison that the efficacy of political technology was less dependent on political culture than had been supposed. Montesquieu was a Bacon, Madison wrote in 1792, but no Newton or Locke; writing "before these subjects were illuminated by the events and discussion which distinguish a very recent period," Montesquieu exaggerated republican government's dependence on civic virtue, falling short of the discovery that an extended republic, properly contrived, could uphold *both* personal and political liberty.[36] Under the Constitution, Madison claimed, the United States was "without a model," virtually a prototype of the technological republic.[37]

Schooled in the civic catechism, most of us will immediately think of the system of checks and balances that Madison described as the "internal" control on government.[38] And as Madison's vaunt suggests, the Constitution involves two departures from established ideas on the separation of powers. In the first place, the legislature is not based on the classes or orders of society. The Framers expected that the Senate's dignity and longer term would attract men of wealth and position, but the Constitution sets no requirement of property or station. The differences between the two houses are purely formal—terms, constituencies, and mode of election—which is what Patrick Henry meant when he sniffed at the "paper checks" of "a sort of mathematical Government" with no necessary connection to social interests.[39] Second, the Framers departed from a strict separation of powers, giving each branch significant power in the affairs of every

34. Conceding that it was important to "annihilate" the states, Gouverneur Morris expressed the hope that the convention would "take the teeth out of the serpents" (James Madison, *Notes of Debates in the Federal Convention of 1787* [Athens: Ohio University Press, 1966], p. 241). On trial by jury in civil cases, see *Federalist* No. 83.

35. *Federalist* No. 1; see also Epstein, *Political Theory of "The Federalist,"* p. 112.

36. *National Gazette*, February 20, 1792, in *Papers of James Madison*, ed. Robert Rutland (Charlottesville: University of Virginia Press, 1983), 14:233.

37. Letter to Edward Livingston, April 17, 1824, in *Writings of James Madison*, 9:188.

38. *Federalist* No. 51.

39. Herbert J. Storing, ed., *The Complete Anti-Federalist* (Chicago: University of Chicago Press, 1981), 5:233, 236.

other, increasing the occasions for contention. As David Epstein points out, the American Framers regarded the separation of powers as a set of devices shaped to avoid the concentration of power that is the aim of human beings in general and public people in particular. The clash between the branches of government in the American system is intended to derive largely from the competition between individuals ambitious for prominence and power, collected in institutions jealous of their prerogatives, a mechanical safeguard against the dependence of private citizens on the politically minded.[40]

This use of rival interests as a substitute for "better motives" is not confined to government: as Madison suggested in the *Federalist*, it is carried into the "whole system of human affairs, private as well as public," where it helps to solve the great riddle of republican politics, the problem of majority faction.[41] Popular habits made republican government a practical necessity in America, but the Framers were also convinced republicans in theory, since majority rule approximates the natural equality and "consent of the governed," which they saw as the origin of all legitimate regimes.[42] Moreover, the "republican principle" follows directly from the determination to avoid an evaluation of souls or ruling qualities. Majority rule makes no judgment on these matters. Given a rule of suffrage and reasonably honest elections, it settles government on a quantitative and objective basis: majoritarianism moves from the rule of speech to the rule of number.

Nevertheless, this solution is delicate. The "just powers" of government may derive from consent, but its *effective* power depends on force or energy. Only in the state of nature is majority rule also the rule of force; in civil society the command of armed forces, wealth, or strategic position give unequal advantages, so that force is ordinarily on the side of minorities.[43] The "republican principle" is a legal bar to minority rule, but if majorities are factional, pursuing a narrow idea of self-interest, there is every danger that strong minorities will refuse to accept the results of elections and appeal to a test of strength.

The Framers' answer, of course, is the large republic, its "orbit" expanded by representation to take in a bewildering variety of inter-

40. Epstein, *Political Theory of "The Federalist,"* p. 141.

41. *Federalist* No. 51.

42. Martin Diamond, "Democracy and *The Federalist:* A Reconsideration of the Framers' Intent," *American Political Science Review* 53 (1959), 52–68.

43. Locke, *Two Treatises of Government,* II, para. 96, 137.

ests and factions.[44] Majorities in such a regime are bound to be shifting coalitions, composed of mutually distrustful groups and
conflicting interests. Evidently, such majorities are assembled
through bargains and compromises and are unable to strike too high
a moral tone. The mediocrity of majorities teaches citizens to limit
their political enthusiasms and allegiances. In a large republic, any
group or party whose program we wholly support, and in which our
views matter, will almost certainly be too small to succeed. An extended republic tends to produce a limited-liability politics in which
moderation is the fruit of disaffiliation and disenchantment.

In fact, in the Framers' thinking, detachment is a kind of substitute
for civic spirit. All attachments are suspect, since the bonds of love
and community limit liberty, tying individuals to particular persons,
places, ideas, and institutions without regard to their usefulness.
Worse, attachments ordinarily chain us to the past, since they are
likely to result from early education and long familiarity, just as ancient examples and teachings—most obviously, religion—tend to
double the strength of opinion. Even the reasoning associated with
political attachment is dangerous and unreliable. "The reason of
man," Madison wrote, "like man himself, is timid and cautious when
left alone, and acquires firmness and confidence in proportion to the
number with which it is associated."[45] Bold reason is the quality of
human beings in a state of association, and—especially in the "noblest minds," driven by the passion for fame—it can enter into "extensive and arduous enterprises for the public benefit."[46] And on rare
occasions like the American Revolution, when private passions are
disciplined by threat and indignation, ordinary citizens may become
a fraternal public capable of heroism and virtue. Such unusual excellence, however, is not the foundation for "a system we wish to last
for ages."[47]

In the ordinary course of things, individuals are more apt to be
rational in isolation. Statesmen and teachers who are visible and accountable may be able to govern passion. For most citizens, however,
the combination of personal anonymity with strength of numbers
tends to unite private passion with public daring, an invitation to

44. *Federalist* Nos. 10, 14, 51.
45. Ibid., No. 49.
46. Ibid., No. 72.
47. *Notes of Debates*, p. 194. In *Federalist* No. 49, Madison listed the conditions at the time
of the Revolution that ensured successful public constitution-making, observing that the
United States could not expect any "equivalent security" in the future.

faction and party spirit. The Athenian assembly would have been a mob, Madison asserted, even if every citizen had been a Socrates.[48] Whatever is gained in human excellence by a fraternal politics is purchased at too high a price. The "bright talents and exalted endowments" produced by the ancient city-states, Hamilton claimed, have only "a transient and fleeting brilliancy," tarnished by the "vices of government" that helped to produce them.[49]

By contrast, the Framers hoped that the design of the laws would weaken attachments, setting individuals free but leaving them somewhat isolated in psychological terms, confronting within civil life a mild version of the order-inducing vulnerability of the state of nature. The reasoning of human beings who are "left alone" is timid, not dazzling, but it can slow even great passions by its very fearfulness, edging the desires toward safe, if petty, objects. And timid reason develops "circumspection," the effort to seek safety by seeing things from all sides, which, while not truly public-spirited, is still public-regarding.

This strategy of detachment is evident in the Constitution's provisions with respect to the states. In the Framers' view, the states, like all political societies, were no more than artifacts intended to serve the purposes of individuals. James Wilson told the Constitutional Convention that "a private citizen of a state is indifferent whether power is exercised by the General or State Legislatures, provided it be exercised most for his happiness." Madison agreed: "The people would not be less free as members of one great republic than as members of thirteen small ones."[50] Yet a citizen of a small republic does enjoy a better chance to be heard or to hold office, and to that extent can be said to have more political freedom. The freedom that concerned Madison and Wilson was the liberty of private persons, and it is by that standard that they preferred to judge the states.

In practice, however, the states could not be evaluated in strictly utilitarian terms. They enjoyed old loyalties—"antecedent propensities"—and were emotionally close to their citizens, the beneficiaries of "the first and most natural attachment of the people."[51] The state governments, consequently, had an interest in established ways and in parochiality, and since "all public bodies" are ruled by avarice and

48. *Federalist* No. 55.
49. Ibid., No. 9.
50. *Notes of Debates*, pp. 162, 166.
51. *Federalist* No. 46.

ambition, the state regimes sought to hold their citizens to narrow and outdated interests, using attachments to limit liberty.[52]

The government of the Union, consequently, is given direct access to individuals within its sphere of authority, without the need for any "intermediate legislation" by the states, in order to make a claim on "those passions which have the strongest influence upon the human heart."[53] At a distance from the bodies, senses, and day-to-day lives of citizens, the central government will always be at a disadvantage as far as affection is concerned.[54] The Framers were confident, however, that interest is naturally sovereign unless confused and opposed by overwhelming attachment. By ending the states' monopoly over "those channels and currents in which the passions of mankind naturally flow," the national government allows interest to make itself felt, eroding state attachments over the long term. Since the federal government can be expected to have a "better administration"—combined with sovereignty over commerce, money, and war—it will gradually detach affections from the states.[55] To be sure, the Union will attract only diffuse affections and relatively weak attachments, but this tepid patriotism, the counterpart of timid reason, suited the Framers' hope of holding government in the role of servant, securing liberty for individuals.

In this version of political hydraulics, commerce is a millrace. Acquisitiveness is the great civilizing passion—it was the "taste for property," Gouverneur Morris argued, that prompted human beings to leave the state of nature—and the new national market opens new opportunities for specialization and gain, liberty and power.[56] Commerce also broadens, gentling some of the self-centeredness of avarice and forcing consideration of other interests.[57] Moreover, since values vary with supply and demand, commercial life promotes detachment and flexibility, an emotional distance from any particular set of products, techniques, and relationships, especially a sensitivity to changes in opinion.

At the same time, the Framers regarded economics as essentially

---

52. *Notes of Debates,* p. 131; see the criticisms of the state regimes by Wilson and Morris, pp. 90, 241.

53. *Federalist* No. 16; see also *Notes of Debates,* p. 74.

54. *Federalist* No. 17.

55. Ibid., No. 27; see also Nos. 17, 34.

56. *Notes of Debates,* p. 244.

57. Hence Hamilton's argument that the "extensive inquiry and observation" of merchants makes them the natural patrons of manufacturers and tradesmen (*Federalist* No. 35).

political, a set of contrivances to be fitted to human wants.[58] The chief "task of modern legislation," Madison observed, lies in regulating the diverse kinds of property.[59] Inequality resulting from differences in ability is acceptable and even desirable since the protection of differing faculties is government's first aim. But wealth will always be tempted to try to shut out competition, just as social and political power can be used to hold others in a disadvantaged position.[60] Commerce should consequently be governed to ensure a reasonable opportunity for all. On the whole, the Framers trusted that the advantage of the large republic—the competition of many interests, denying more than short-lived ascendancy to any—would be effective in economics as well as politics, allowing regulation to be left largely to "the silent operation of the laws."[61]

Yet despite their confidence in political technology, the Framers recognized that a republican people needs some element of moral and civic virtue, a "vigilant and manly spirit" that "nourishes freedom."[62] Self-preservation and self-interest, for example, do not support military virtue; in defense of a standing army, Hamilton observed that a commercial people can never be a "nation of soldiers."[63] But unless citizens are patriotic enough to defend their liberties, a republic is wholly dependent on and vulnerable to an army of mercenaries or dedicated soldiers, employees who are doubly dangerous since they are apt to despise the soft and ignoble values of a liberal regime.

The problem of the military is only an instance of a more basic problem. It is clearly in my interest for a political society to exist and for other citizens to obey the law, but it also appears to be in my interest to break the law whenever it seems advantageous, enjoying the benefits of law without being limited by it. The Framers' doctrine can only answer that, in practice, lawlessness risks too much: I am likely to be found out and punished; moreover, any crimes that I commit endanger the social contract and political society, especially if I escape punishment, because they weaken the trust and fidelity

58. In 1729, for example, Franklin argued that the government should manage the money supply in the interest of general economic prosperity (*Benjamin Franklin: Writings*, pp. 119–35).

59. *Federalist* No. 10.

60. *Papers of James Madison*, 9:76.

61. *Writings of James Madison*, 6:86.

62. *Federalist* No. 57.

63. Ibid., No. 8.

of others. This argument is plausible in the case of the great and powerful, who are too prominent to escape notice and whose wrong-doings may seem to justify others in disregarding the law. For private persons, however, these dangers are greatly diminished; the obscure and unpropertied have a good chance of escaping detection, while their conduct is less likely to set a destructive example. The poor and desperate, moreover, may feel that they have nothing to lose. Such people, the Framers observed, need religion and moral education to make good the deficiencies of private interest and public power as supports for law.[64]

In these terms, however, moral education is a kind of equivocal fabling, practiced on behalf of the "aggregate interests of the commu-nity" on individuals who are unable to reason or to discern their interests (that is, according to the Framers, children and perhaps women), or whose very interests may nudge them in a lawless direc-tion, such as "desperate debtors" like Daniel Shays and, even more clearly, slaves.[65] It is not true that promises should always be kept, but it is useful for people to believe so; not all crimes are detected and punished here below, and a belief in a Supreme Judge sustains many who might otherwise falter.[66]

Moral education, in other words, is necessary but vaguely ques-tionable, too dangerous to be trusted to the state and also incompat-ible with the dignity of a regime devoted to reason and freedom. At the same time, there are distinct problems in entrusting moral educa-tion to the private sphere or to the states. Lacking a proper founda-tion in reasoned consent, early education's authority and moral education's deceptions can be justified only if they develop or support the individual's capacities for independence and reason. The civic foundation of the technological republic requires an appropriate "education for liberty."[67]

John Locke and his American followers supplied the founding gen-eration with just such an educational technology, one distrustful of

64. Franklin argued for the usefulness of a "Publick Religion," particularly given "the Advantage of a religious Character among private Persons" (*Benjamin Franklin: Writings,* p. 336).

65. Hamilton called Shays a desperate debtor in *Federalist* No. 6, probably alluding to *Leviathan,* chap. 11.

66. A belief in "particular Providence," Franklin held, is necessary to "weak and ignorant Men and Women, and . . . inexperienced and inconsiderate Youth of both Sexes" (*Benjamin Franklin: Writings,* p. 748).

67. Nathan Tarcov, *Locke's Education for Liberty* (Chicago: University of Chicago Press, 1984).

authority and attachment, the mirror of the Framers' political principles.[68] In Locke's view, children, like all human beings, are driven
by the desire for freedom and mastery, and hence are naturally inclined to resist commands and overt rule. Education, accordingly, is
best understood as a problem in human hydraulics: once their nature
is understood, children are "flexible waters" whose courses can be
channeled, so that by rightly designed training, even defects of nature can be at least "a little remedied."[69]

In the American tradition, this emphasis on malleability has often
been exaggerated into a belief in the almost limitless power of "socialization." Locke's view was soberer, since he regarded nature as a limit
on education. Nevertheless, he did teach that apart from certain basic
impulses, such as hunger and the yearning for liberty, human desires
have no natural objects and can be directed by proper technique,
so that, for example, play can be made burdensome for children by
imposing it as a duty, making study relatively more attractive.[70]

But Locke also sets out to *promote* a flexible soul that will be respon-
sive to socialization and to opinion through a family government that
is severe, to say the least. He argues that parents must attempt to
discipline children's minds and habits from their "very Cradles,"
when they are "most tender, most easy to be bowed," and especially
"before Children have Memories to retain the beginnings of it."[71] This
intervention is to be stern: apart from satisfying a child's physical
needs, Spartanly defined, parents must make a special point of denying children what they want and must also ignore crying except when
a children has suffered some bodily hurt.[72]

This program has two aims. First, it is intended to make the habit
of self-denial and self-control second nature. The desires aim at natural goals (in fact, there are times when it may be necessary to stimulate them), but the ability to defer gratification is necessary for great
or long-term projects and for the defense of liberty, which often calls

68. Locke and his school, Daniel Calhoun writes, addressed "the crucial problem of technique" (*The Intelligence of a People* [Princeton, N.J.: Princeton University Press, 1973], p. 154).

69. Locke, *Some Thoughts concerning Education,* para. 66, 73, 87; Tarcov, *Locke's Education,*
pp. 82, 114.

70. In fact, Locke held that education is responsible for nine-tenths of character (Tarcov,
Locke's Education, p. 82). For Locke's argument on play and study, see *Some Thoughts concerning
Education,* para. 128.

71. *Some Thoughts concerning Education,* para. 33–35, 38, 44.

72. Ibid., para. 40, 78, 111–13; Locke also warns against excessive strictness, since beyond
a certain point it only stimulates desire (para. 20).

for sacrifice and pain.[73] Second, Locke's program for early education is designed to avoid the fostering of dependence. It teaches children that they cannot simply or safely rely on others, a negative spur toward managing for themselves. And because, on this view, women are too prone to indulgence, the role of the mother in early child-rearing should be diminished, overruled by the government of the father, the voice of civil reason overriding natural affection.[74]

At the same time, the frustrations children suffer under Locke's scheme are intended to drive home the lesson of their weakness and practical vulnerability. Parents are to curb their children's desire for dominion and to break their pride, brooking no "Obstinacy or Rebellion," using milder modes if possible but, if necessary, falling back on force.[75]

These formative severities reproduce, psychologically, the experience of the state of nature, leaving children free but driving them into society. Forgotten, but written in the secret places of the soul, early harshness reduces the need for external authority and physical constraint later in life. In Locke's system, rule is replaced by repression.[76] Detached but sensitive to the power of others, Locke's citizen is intended to develop a "relish" for "the Good Opinion of Others," and with it the polite circumspection the Framers valued.[77] Classical education aimed at souls who love what is true more than what is honored, a virtue defined by inner quality rather than outward behavior. Locke's standard is lower, but more accessible, and less dangerous to civil peace than pride. The Lockean school aims to rear civil competitors whose desire for supremacy is defined in terms approved by society and by law and whose very self-interest is bent toward generosity—liberal souls for a liberal regime.[78]

There were evident problems, however, in translating Locke's prescriptions into an American idiom. He wrote primarily of an education that, after early childhood, would be entrusted to tutors, but

73. Ibid., para. 45, 126; Tarcov, *Locke's Education*, pp. 86, 90, 176–77.

74. *Some Thoughts concerning Education*, para. 4, 7, 13, 34, 216; Locke's view that nature assigns responsibility to both parents (*Two Treatises of Government*, II, para. 53) will sound better to modern ears.

75. *Some Thoughts concerning Education*, para. 78, 84, 87, 104, 106–8, 112.

76. Calhoun, *Intelligence of a People*, pp. 143, 154; *Some Thoughts concerning Education*, para. 43, 44.

77. *Some Thoughts concerning Education*, para. 56, 58, 62, 143, 200; Tarcov, *Locke's Education*, pp. 101–41.

78. *Some Thoughts concerning Education*, para. 109; Tarcov, *Locke's Education*, p. 141.

such training was out of reach for most Americans, and even schooling still seemed beyond the needs and resources of ordinary citizens.

It was apropos, then, that when Benjamin Franklin resumed his
Autobiography in 1784, he redirected the book from his son to the
public, as if acknowledging that the new republic needed an education in which the family would yield to the polity.[79] The change was
also consistent with Franklin's long-standing hope of developing a
model for self-education within the power of the many, a secular
"apparatus" of moral training, more reliable than Christianity—since
"all Men cannot have Faith in Christ"—and thus "adapted to universal Use."[80] Taming lightning was less ambitious: aiming to improve
on Christianity, Franklin was also aspiring to outdo Locke, in so many
ways his master. A practiced enchanter, Franklin spoke to generations
of Americans with unique authority, and he still casts a long shadow.

Most Americans know Franklin's technology of the virtues, if only
in a somewhat distorted mirror, since it has become one of the hazards of American childhood. He recommended making and retaining
a chart of thirteen virtues; each week the individual is to practice one
of these graces, grading his or her achievement daily to enlist pride
on the side of good performance. After the thirteen weeks of practice,
Franklin confided, individuals will be greatly strengthened in the
habits of virtue, by no means perfect but better and happier.[81]

His method presumes morally serious individuals who desire to
lead good lives "but know not *how* to make the change."[82] The system,
consequently, depends on an adequate early rearing, affording that
"tolerable character" with which Franklin, by his own account, had
begun his career.[83] But Franklin did not recommend his own rearing
to others, substituting a more modern education based on his adult
experience.

He "often regretted," Franklin tells us, that he devoted much of
his early reading to his father's books of "polemic Divinity," but he
has kinder things to say about Bunyan's *Pilgrim's Progress*, Defoe's
Essay on Projects, Mather's *Essays to Do Good*, and Plutarch's *Lives*. His

79. Franklin, *Autobiography*, p. 77.

80. Ibid., p. 80; letter to Lord Kames, May 3, 1760, in *Benjamin Franklin: Writings*, p. 766.

81. Franklin, *Autobiography*, pp. 91–103; Franklin also developed a "Moral or Prudential
Algebra," in his letter to Joseph Priestley, September 19, 1772, in *Benjamin Franklin: Writings*,
pp. 877–78.

82. Letter to Lord Kames, May 3, 1760, in *Benjamin Franklin: Writings*, p. 765; see also
Autobiography, p. 100.

83. *Autobiography*, p. 64.

bow to Mather is often the object of comment, but Franklin notes only that Mather's book "perhaps" gave him a "Turn of Thinking that had an influence" on his later life. Similarly, while expressing no regret at reading Bunyan and Defoe, he has no praise for their teachings, only for their prose styles, which he subsequently lauds. By contrast, reading Plutarch—the only pagan text on Franklin's list— was "time spent to great advantage," a lesson underlined by his decision to sell Bunyan to buy histories, the progress of the soul giving way to progress in time. And as that might suggest, even Plutarch's wisdom proved inferior to the "peaceful, acquiescing manners" of Franklin's teaching: ancient virtues must also yield to modern experience and science.[84]

For Franklin, the heroic virtues, anciently praised, have no just claim to rule. The love of honor provokes contention and violence, so that warriors and statesmen ("public cutthroats and intriguers"), together with contemplatives and poets, have failed to pacify existence, a goal that requires the reconstruction of human nature.[85] Philosophy is a somewhat better guide, since Franklin admired Socrates' guileful rhetoric, his pose of "the humble Inquirer and Doubter." But by confuting one's opponents, one provokes them; skill in debate brings victories but not good will.[86] Ultimately, Socratic rhetoric is not safe because it teaches too openly and too much in public. Instruction and mastery must be more thoroughly hidden.[87] Speech— for Aristotle, the political quality par excellence—is less effective than experiments that others can "repeat and verify" and need no verbal defense.[88] The model of government is not the Bible's God, who creates by speech, but the silent deity who orders inanimate nature through the laws of attraction and motion.

"Publick Religion," for example, should be devoted to morals rather than faith or doctrine, and preaching should aim to make "good Citizens" at least as much as it tries to make "good Presbyterians."[89] Stripped to its "essentials," religion is reduced to a belief in God, the

84. Ibid., pp. 13, 24, 81, 83.

85. Ibid., pp. 81, 83; letter to Joseph Priestley, June 7, 1782, in *Benjamin Franklin: Writings*, pp. 1047–48.

86. *Autobiography*, pp. 18, 147–48.

87. As Ralph Lerner points out, when Franklin appeals to Pope's authority to support this argument, he underlines the point by improving on Pope's poetry ("Benjamin Franklin: Spectator," in *The Thinking Revolutionary* [Ithaca: Cornell University Press, 1987], pp. 51, 53).

88. *Autobiography*, pp. 173–74; Aristotle, *Politics*, 1253A8–1253A18.

89. *Autobiography*, p. 89; *Benjamin Franklin: Writings*, p. 748.

immortality of the soul, and the due reward of vice and virtue "either here or hereafter."[90] This faith lacks any Christian dimension, even the ambiguous affirmations of Locke's *Reasonableness of Christianity*. A Maker and Creator, Franklin's God does not redeem the world, except possibly in the next life; here below, He supports the quest for mastery, so that it may be "the Design of Providence" to extirpate the Indians "to make room for Cultivators of the Earth."[91]

Humility, that high Christian excellence, was included in Franklin's list of virtues but as an afterthought, added only at the urging of a Quaker friend. Yet humility did not fit: a desolating sense of one's own shortcomings combined with wonder at God's glory, humility is a state of the soul as paradoxical as Socrates' knowledge of ignorance, the gift of grace and not human teaching. As Franklin winkingly observed, even if it were possible to cultivate the virtue, success would have made him "proud of [his] humility."[92] Nevertheless, Franklin argued, the *appearance* of humility is of great value, since its disguise allows one to lead those for whom appearances are everything. Since to "raise one's Reputation" is to invite envy and opposition, the truly great teacher or founder conceals himself behind an apparent humility, sustained by an inner pride in his creation that does not need public recognition.[93] Out of sight, it is possible to appeal to and channel the lesser pride of less perfected human beings, holding up the lure of honor and success. For Franklin as for Locke, pride and vanity, the most tenacious of the passions, become the bases of social morality. The desire for approbation can tame selfishness into civil competition, where it can be guided toward benevolence and the common good.[94]

The truly reigning virtue, in Franklin's system, is sincerity, which he makes central in his list of excellences. Unlike more modern versions, Franklin's sincerity is an *inward* quality, compatible with useful trickeries, limited only by the injunction to "use no hurtful deceit.[95]

To be sincere is to be honest *with oneself,* to maintain a directing will at the center of all the social and political roles it may be necessary or useful to play. Like self-mastery, sincerity requires that one never

90. *Autobiography*, p. 89.
91. Ibid., pp. 135–36.
92. Ibid., p. 103.
93. Ibid., pp. 87, 102.
94. Ibid., pp. 4, 113; Paul W. Conner, *Poor Richard's Politics: Benjamin Franklin and His New American Order* (New York: Oxford University Press, 1965).
95. *Autobiography*, p. 92.

identify with such disguises, keeping psychological distance between oneself and one's relationships.

This detached self-command is the motif of the "Party of Virtue" that Franklin suggested, based on a new elite, a secret "Society of the Free and Easy" composed of persons who—having followed Franklin's course of the virtues—would be free from vice and debt and able to set their own directions. The vanguard of liberty, the society was to operate behind the scenes "until it was become considerable."[96] Ordinary political parties, Franklin argued, remain united by common principles only when out of power; once successful, parties tend to fragment, divided by the interests and vanities of their members. His projected party, Franklin thought, would be an exception because its members, practiced in his discipline, would be free from the itch for public esteem. Franklin hoped to school a new class, and perhaps a public, that would be literally utopian, detached from everything except the project of mastering nature—suitable citizens for the technological republic.

Two centuries later, we are all participants in the Framers' project, and the great majority of us appreciate its benefits and cherish its liberties. At the same time, we know that power has proved more difficult to control than the Framers had hoped. The "silent operation of the laws" has failed to avoid monumental inequalities of wealth and power, both of which constrain private opportunity and political freedom. Moreover, citizens—so many parts of a complex division of labor in an integrated market and world—are radically dependent on and exposed to national and international events. As Michael Ignatieff comments, this vulnerability, so thoroughly beyond the control of individuals or their social relationships, tends to transform needs into rights that must be protected by public authority.[97] Even those political scientists who discern a somewhat self-regulating mechanism in the competition among interest groups see a need for government to broker compromises and maintain fair rules of the game, while conservative administrations acknowledge the need for an economic safety net and use public money to protect private savers. The need to curb private inequalities and to provide citizens with minimal security—both necessary to republican life—has created and continues to require a government more active and more inclined to intervene in the private sphere than the one the Framers brought into

96. Ibid., p. 105.
97. Michael Ignatieff, *The Needs of Strangers* (New York: Viking, 1985).

being. Moreover, the growth of power and pace of change demand swift, sometimes secret decisions, so that the president's prerogative strains against, and sometimes overruns, the barriers and forms in which the Framers sought to confine it.[98]

Meanwhile, the republic's citizens—its "human resources"—seem increasingly less suited to control its energies. The Framers trusted that the scale and moderation of public life would limit the confidence and enthusiasm of citizens, turning them toward private life. As Tocqueville taught us, however, mass politics magnifies this design until it becomes grotesque. Contemporary majorities, all but incomprehensible to most citizens, are certainly beyond their power to change. Even public leaders scurry to adapt to opinion, and presidents and their staffs anxiously watch the polls. Private citizens quickly learn that public life is a sphere of weakness and indignity in which they matter only as statistics, and in which they are safe only if they confine their thoughts and voices within the bounds permitted by the "tyranny of the majority."[99] Small wonder that the retreat into private life becomes a rout, or that the republic suffers from a chronic shortage of allegiance.

Private life, however, is no safe citadel. Gradually, American society has been yielded to the laws and logic of technique.[100] The temptations of commerce bribe us, and change forces us, to fear commitments to places or persons, institutions or ideas. Attachments make us liable to pain, loss, and abandonment, just as they are obstacles between us and new opportunities. And individualism, in alliance with the laws, offers justifications and routes for escape.[101]

As families and communities have weakened, they have become more exposed and more apt to conform to the dominant institutions and ideas. Today the technological republic pursues Americans into the most private places and to the very foundations of the self. From Locke to our times, Americans have been attracted to technologies of child-rearing ranging from the sensible to the bizarre. For us, however, the impact of technology grows ever more direct: the media play an increasing role in the education of children, before literacy and even before speech, encouraging the development of spectatorial per-

98. Harvey C. Mansfield, Jr., *Taming the Prince* (New York: Free Press, 1989), especially pp. 1–20.

99. *Democracy in America*, 1: chap. 15.

100. Jacques Ellul, *The Technological Society* (New York: Knopf, 1964).

101. Lewis Lapham, "Notebook: Supply-Side Ethics," *Harper's* (May 1985), 11. Of course Tocqueville anticipated this (*Democracy in America*, 2, bk. 2, chap. 2).

sonalities careful to preserve their flexibility and their distance. That
teaching is reinforced by the discovery, through observation if not
personal experience, that families and friendships are unstable relationships in which it is wise to limit one's liabilities.[102] The sources of
commitment and loyalty are becoming anorexic in the private order
as well as in public life, so that Tocqueville's bleak fear—that in the
end each human being would be confined "in the solitude of his own
heart"—takes on a disturbing immediacy.[103]

Lacking the support of communities of faith and remembrance,
individuals face oblivion as well as insignificance, so it is not surprising that more Americans grasp at immediate or short-term gratifica-
tions, or that they find them hollow.[104] As the sphere of individual
liberty shrinks in practice, Americans seem more insistent on being
unrestrained in what remains. In intellectual life the accurate perception of our dependence on other human beings and on the environment is accompanied by a desperate insistence on the individual's
freedom in theory and in the soul. The very idea of a "self" is widely
regarded as a violation of the freedom of subjects to construct themselves, and even words are asserted to be so many chains imposed
by power.[105]

Yet this rebellious distrust of language mirrors the official doctrine
of technological government, always disposed to substitute rules and
procedures for speech and judgment.[106] In contemporary public life,
this inclination comes to seem a virtual necessity: citizens, disdaining
politics and inadequately informed about it, are ill equipped to deal
with the modern world's intricacies and perils. Power and scale, the
artifacts of technological politics, have magnified its original theory
and practice into the ideology of the Administrative State and the
search for scientific or quantitative measures by which objective data
can displace civic deliberation.[107] Just as power moves from the legisla-

102. Judith Wallerstein, "The Impact of Divorce on Children," *Child Psychiatry* 3 (1980),
459–62.

103. *Democracy in America*, 2, bk. 2, chap. 2.

104. Ibid., chap. 13; bk. 3, chap. 13; Ignatieff, *Needs of Strangers*, pp. 83–103; Russell Jacoby,
Social Amnesia (Boston: Beacon, 1975).

105. Marvin Zetterbaum, "Self and Subjectivity in Political Theory," *Review of Politics* 44
(1982), 59–82.

106. Sheldon S. Wolin, "Democracy in the Discourse of Postmodernism," *Social Research*
57 (1990), 5–30.

107. Wolin, *Politics and Vision*, pp. 352–434; Dwight Waldo, *The Administrative State* (New
York: Ronald, 1946); Calhoun, for example, calls attention to the growth of professional
licensing and testing (*Intelligence of a People*, pp. 319–20).

ture to the executive in the quest for expert decision, it has tended to pass from the executive into the bureaucratic "technostructure," sliding "into the body of the organization," increasingly out of the public's sight and reach.[108]

Even if this pattern reduces many political risks, it also lessens the public's support for government. It convinces a growing number of citizens that public life is a sham, manipulated by secret powers and hidden strategists, at best no answer to and possibly complicit in their own indignity. Resentment, thus far, is largely held in check by perceived weakness and the consolations of affluence, expressed in private fantasy or self-destructiveness, but it is no comfortable ground for the republic's future.[109] The public's indifference to politics and its disposition to indignation—exploited by so many investigative reporters—are warning signs of the despair and rage rooted in the technological project itself. Not only does the mastery of nature entail the possibility of mastering human nature, but the elimination of human need seems to erode the need for what is human—our labor and art as well as our allegiance and aspiration.

Tocqueville described two ways of mitigating, and perhaps healing, the ills of the technological republic. In the first place, he argued for the decisive importance of the "art of associating together" as a protection against the impotence felt by individuals in mass regimes.[110] This craft is not easily learned in modern America: Franklin contrived voluntary associations for street paving and fire protection, but we depend on the public funding of public services. Of course, it *is* possible for us and for the government to seek to expand the sphere of association wherever possible. We could abandon or reverse the hostility to party that has informed our legislation throughout this waning century, or we could follow the Wagner Act's more positive example, encouraging association by the support of law. In the same way, public policy can attempt to do more to assist families and communities, acting—against the teaching of the laws—on the principle that strong and secure attachments are the foundations of self-esteem and freedom.[111]

Any such policies, however, also point to the need—one Tocque-

---

108. John Kenneth Galbraith, *The New Industrial State* (Boston: Houghton Mifflin, 1967), and "Coolidge, Carter, Bush, Reagan," *New York Times*, December 12, 1988, A16.
109. Robert Coles, "The Politics of Ressentiment," *New Republic*, August 2, 1982, 32–34.
110. *Democracy in America*, 1, chap. 12; 2, bk. 2, chap. 7.
111. Robert Karen, "Becoming Attached," *Atlantic*, February 1990, 35–39.

ville observed—for "constituted authorities" to support "theoretical studies," allowing, and if necessary compelling, Americans to consider the truth that is beyond possessions and beyond associations, the nature in which humans and political societies have their being. The artist Jacques Louis David was inferior to Raphael, Tocqueville commented, even though he followed nature like a good anatomist, capturing the details of the human body. Like technology, David "dragged nature to the easel," attempting to know it through its parts and to master it by force. Raphael, by contrast, drew from the inside out, from first principles rather than appearances, informed by a sense of the whole that gives his work greater "life, truth and freedom."[112] Compare this trinity of Tocqueville's with Locke's: truth deprives liberty of the central position, while property is displaced altogether. In the most fundamental sense, Tocqueville implies, freedom is not the mastery of persons and things; it is being what we are, subject to truth's authority. No teaching is more necessary if the technological republic is to rediscover its soul.

112. *Democracy in America*, 2, bk. 1, chaps. 10, 11. In the text of chap. 11, Tocqueville comments that Raphael's aspiration to "surpass nature" made him less concerned with rigorous accuracy. In a footnote, however, he argues that Raphael's drawings display a superior "fidelity to nature" and "profound scientific knowledge."

Suggestions for Further Reading

Adair, Douglass. *Fame and the Founding Fathers.* New York: North, 1974.
Horwitz, Robert, Ed. *Moral Foundations of the American Republic,* 3d ed. Charlottesville: University Press of Virginia, 1986.
Kasson, John. *Civilizing the Machine: Technology and Republican Values in America,* 1776–1900. New York: Grossman, 1976.
Lienesch, Michael. *New Order of the Ages: Time, the Constitution, and the Making of American Political Thought.* Princeton, N.J.: Princeton University Press, 1988.
Pangle, Thomas. *The Spirit of Modern Republicanism: The Moral Vision of the American Founders and the Philosophy of Locke.* Chicago: University of Chicago Press, 1988.
Weinberger, Jerry. *Science, Faith, and Politics: Francis Bacon and the Utopian Roots of the Modern Age.* Ithaca: Cornell University Press, 1985.

5 Romanticism and Technology: Satanic Verses and Satanic Mills

PAUL A. CANTOR

> And Science, and her sister Poesy,
> Shall clothe in light the fields and cities of the free!
> —PERCY SHELLEY, *The Revolt of Islam*

I

We all think we already know what the Romantics thought about technology: they hated it. Perhaps the most famous response to technology in English Romantic poetry are these lines from the Preface to Blake's *Milton:*

> And did those feet in ancient time,
> Walk upon Englands mountains green:
> And was the holy Lamb of God,
> On Englands pleasant pastures seen!
>
> And did the Countenance divine,
> Shine forth upon our clouded hills?
> And was Jerusalem builded here,
> Among these dark Satanic Mills?[1]

---

1. *Milton,* pl. 2, ll. 1–8. All quotations from Blake are taken from *The Complete Poetry and Prose of William Blake,* ed. David V. Erdman (Berkeley: University of California Press, 1982).

109

Though some scholars have questioned whether Blake had industrial mills specifically in mind here,[2] in the common understanding of these verses Blake views the so-called technological progress in England as an act of desecration. Before the coming of the Industrial Revolution, the English landscape was a natural paradise, with "mountains green" and "pleasant pastures." Technology has clouded over this natural beauty, substituting man-made horrors for the divine world in which we were originally intended to live. What we view as Romantic in this poem is its apparent piety to nature; it has become a critical commonplace that the Romantics rejected technology as an unnatural transformation of the human condition.

Like all critical commonplaces, this understanding of Romanticism obviously has some basis in reality. Wordsworth seems to share this reverence toward nature, as well as the hostility to the modern scientific attitudes that have led to technological change. It is, after all, Wordsworth who wrote the line that brilliantly sums up all that is dubious in the scientific/technological approach to the world: "We murder to dissect."[3] But he also wrote a sonnet with the improbable title "Steamboats, Viaducts, and Railways":

> Motion and Means, on land and sea at war
> With old poetic feeling, not for this,
> Shall ye, by Poets even, be judged amiss!
> Nor shall your presence, howsoe'er it mar
> The loveliness of nature, prove a bar
> To the Mind's gaining that prophetic sense
> Of future change, that point of vision, whence
> May be discovered what in soul ye are.
> In spite of all that beauty may disown
> In your harsh features, Nature doth embrace
> Her lawful offspring in Man's art; and Time,
> Pleased with your triumphs o'er his brother Space,
> Accepts from your bold hands the proffered crown
> Of hope, and smiles on you with cheer sublime.[4]

Much in this poem seems compatible with Wordsworth's usual atti-

2. See, for example, Harold Bloom's note on this passage in Erdman, *Complete Poetry of Blake,* p. 910.

3. See "The Tables Turned." For a similar view of science in Keats, see *Lamia,* pt. 2, ll. 229–37.

4. All quotations from Wordsworth are taken from *William Wordsworth: The Poems,* ed. John O. Hayden (Harmondsworth: Penguin Books, 1977).

tude toward technology. He still speaks of technological developments as "at war/With old poetic feeling" and charges that they "mar/
The loveliness of Nature." But something has obviously altered when
Wordsworth nevertheless insists that "Nature doth embrace/Her lawful offspring in Man's art."

Perhaps the clue to his change of heart is to be found in the fact
that Wordsworth speaks of technology as "at war with *old* poetic
feeling." It has evidently begun to dawn on him that hostility to
technology just might land him in the camp of the old fogies. By
associating technological developments with a "prophetic sense/Of
future change," he is admitting that technology may be the wave of
the future, and apparently he does not want to be left behind in its
wake. Above all, Wordsworth comes to recognize a triumph of human
power in modern technology. It may at first be difficult to accept the
fact that the Wordsworth whom we normally expect to be worshiping
nature appears in this sonnet to be worshiping speed like a wide-
eyed youngster at a drag strip. The sonnet even ends with one of the
central operative words of Romantic discourse: *sublime.* In "Steamboats, Viaducts, and Railways," Wordsworth makes his peace with
modern technology insofar as he finds in it evidence of the sublimity
of the human spirit, its ability to triumph over and transcend the
natural world.

As atypical as this sonnet may be, it is still a useful reminder of
some of the internal contradictions in Romanticism that prevented
the authors from having a simple and straightforward response to
technology. The Romantics' premium on novelty and innovation was
at odds with their hostility to the advance of technology, arguably
the most progressive force of their era. How indeed could this revolutionary generation entirely reject the Industrial Revolution? In fact,
the great manifesto of the Romantic revolution in England—Wordsworth's Preface to the *Lyrical Ballads*—contains a seldom noted and
again seemingly uncharacteristic passage in which he connects his
notion of poetry with modern science, and it is the concept of revolution that supplies the link:

> If the labours of Men of Science should ever create any material
> revolution . . . in our condition, . . . the Poet will sleep then no
> more than at present, but he will be ready to follow the steps of
> the Man of Science. . . . The remotest discoveries of the Chemist,
> the Botanist, or Mineralogist, will be as proper objects of the Poet's

art as any upon which it can be employed. . . . If the time should ever come when what is now called Science, thus familiarized to men, shall be ready to put on, as it were, a form of flesh and blood, the Poet will lend his divine spirit to aid the transfiguration.[5]

In what amounts to a prophecy of the genre of science fiction, which was, in fact, inaugurated in the Romantic era, Wordsworth shows that science and literature were not simply opposing forces for Romanticism. Though the aesthetic sensibility of the Romantics was often offended and repelled by technological developments in their day, they at the same time had to recognize that science and technology had become forces for changing the world. And since this was a goal of the Romantics themselves, they could not afford to ignore or simply scorn technology.

Percy Shelley, for example, was fascinated by modern scientific and technological developments. He dabbled in chemistry, and as early as his student days at Oxford, he had high hopes that technology could solve many of the problems of humanity.[6] In his *Defence of Poetry*, an essay provoked by an argument that science was replacing literature in the nineteenth century, Shelley counters by claiming that Francis Bacon should be numbered among the great poets of humanity—Bacon, the supreme advocate of modern technology.[7] If this seems an aberration, I would point to one of the culminating moments of English Romanticism, the end of William Blake's *Jerusalem*. There he outdoes Shelley by portraying a scene in which "Bacon &

5. Ibid., 1:881–82.

6. In a memoir of Shelley's life, his college friend Thomas Jefferson Hogg reports long conversations about scientific and technological progress. Consider, for example, this reported statement by Shelley: "Water, like the atmospheric air, is compounded of certain gases: in the progress of scientific discovery a simple and sure method of manufacturing the useful fluid . . . may be detected; the arid deserts of Africa may then be refreshed by a copious supply, and may be transformed at once into rich meadows, and vast fields of maize and rice." Shelley's social concerns fuel his hopes for technology: "What a comfort would it be to the poor . . . if we could furnish them with a competent supply of heat!" The new discovery of electricity particularly fired Shelley's imagination: "What a mighty instrument would electricity be in the hands of him who knew how to wield it, in what manner to direct its omnipotent energies; and we may command an indefinite quantity of the fluid; by means of electrical kites we may draw down the lightning from heaven! What a terrible organ would the supernal shock prove, if we were able to guide it; how many of the secrets of nature would such a stupendous force unlock!" For these quotations and similar ones, see Thomas Jefferson Hogg, *The Life of Percy Bysshe Shelley* (London: George Routledge, 1906), pp. 51–52.

7. See *Shelley's Poetry and Prose*, ed. Donald Reiman and Sharon Powers (New York: Norton, 1977), p. 502. All quotations from Shelley are taken from this edition. The essay Shelley was responding to in the *Defence* is Thomas Love Peacock's "Four Ages of Poetry."

Newton & Locke" are reconciled with "Milton & Shakspear & Chau-
cer" (pl. 98, l. 9) in a grand apocalyptic gesture that signals the overcoming of the fallen world. The equally apocalyptic Night the Ninth
of Blake's *Four Zoas* ends with the line "The dark Religions are departed & sweet Science reigns" (p. 138, l. 10). These words may be
less in the spirit of the Enlightenment than Alexander Pope's more
famous "God said, *Let Newton be!* and All was *Light*," but they certainly do sound strange coming from a poet normally thought to be
irreconcilably hostile to science.

I am trying to reopen the question of the relation of Romanticism
to the Enlightenment. In the standard view of literary history, the
Romantics are said to be reacting to the Enlightenment, in particular,
rejecting its overemphasis on rationality.[8] From a narrow historical
perspective, this view is no doubt correct. The Romantics did reject,
often vehemently, their eighteenth-century predecessors, and one
can certainly find many anti-Enlightenment statements by the Romantics, especially Blake (such as his poem "Mock on Mock on Voltaire Rousseau"). But from a broader historical perspective, I want to
argue that Romanticism was, in fact, a continuation of the Enlightenment, or at least did not signal a radical break with it. That is to
say, the Romantics criticized the Enlightenment only on the basis of
principles the Enlightenment itself had developed. As indicated by
Blake's line "The dark Religions are departed & sweet Science reigns,"
the Romantics accepted and welcomed the fundamental achievements of the Enlightenment: the emancipation of humanity from centuries of superstition and the establishment of the autonomy of the
individual, achievements in which the advance of modern science
and technology played a crucial role. Despite their various flirtations
with medievalism, the Romantics never genuinely wished humanity
to return to the conditions of the Middle Ages. They may have questioned the historical outcome of the French Revolution, but they accepted the ideals it tried to put into practice, above all, the goals of
liberty and equality for all. The Romantics' attraction to the Middle
Ages reflects their sense that the Enlightenment's emphasis on individual autonomy had undermined the equally legitimate human
value of community.

Thus Romanticism can be understood as an attempt to correct the

---

8. For a recent restatement of the view that the Romantics rejected the Enlightenment, see
Fred Wilson, "Wordsworth and the Culture of Science," *Centennial Review* 33 (1989), 322–92.

Enlightenment on the basis of Rousseau's criticism of it. But there is a difference between Rousseau and Romanticism.[9] Rousseau saw genuine antinomies in the human condition. Above all, he did not believe that the goals of freedom and community could ever be wholly reconciled. The Romantics were more romantic. They believed it possible to preserve whatever was positive in the heritage of the Enlightenment while rejecting whatever was negative, thereby achieving an ideal synthesis of human values. This tendency is nowhere more evident than in their attitude toward technology. To be sure, they worried about the effects of technological development, particularly the way machinery was tending to dehumanize the European way of life. But, as I have shown, the Romantics also sensed that technology was a new source of human power, which might in true Baconian fashion be applied to "the relief of man's estate." The Romantics hoped to humanize modern science and technology, to enlist them in the service of their utopian vision of the future of humanity.

For example, in another of the great apocalyptic visions of Romanticism, *Prometheus Unbound*, Shelley fills the poem with references to contemporary scientific developments in fields as diverse as paleontology and astronomy.[10] His culminating vision of the redemption of humanity rests on the dethroning of traditional and inhibiting images of divinity once we learn to overcome our alienation from nature and to appropriate the powers of the universe as our own. In Act IV, a vocal Earth confesses to the triumph of the spirit of man over all nature, as scientific exploration solves all mysteries:

> The Lightning is his slave; Heaven's utmost deep
> Gives up her stars, and like a flock of sheep
> They pass before his eye, are numbered, and roll on!
> The Tempest is his steed,—he strides the air;
> And the abyss shouts from her depth laid bare,
> "Heaven hast thou secrets? Man unveils me, I have none."
>   (IV.418–23)

9. For a detailed discussion of this difference, see my book *Creature and Creator: Mythmaking and English Romanticism* (Cambridge: Cambridge University Press, 1984), especially pp. 20–23.

10. For a detailed discussion of this point, see Carl Grabo, *A Newton among Poets: Shelley's Use of Science in Prometheus Unbound* (Chapel Hill: University of North Carolina Press, 1930). See also Alfred North Whitehead, *Science and the Modern World* (New York: Macmillan, 1925), pp. 84–85, especially Whitehead's suggestion that "if Shelley had been born a hundred years later, the twentieth century would have seen a Newton among chemists."

Somewhat more abstractly, Shelley is here celebrating the same triumph of time over space that Wordsworth proclaims in his "Steamboats" sonnet, going the older poet one better by imagining human
flight, and perhaps even space travel. For Shelley, the triumph of
humanity is a triumph of human power over nature, which in the
end yields all its secrets to the inquiring scientific mind. It is no
accident that the great prophet of modern science, H. G. Wells, was
deeply influenced by Shelley and even chose to rewrite *Prometheus
Unbound* in the form of a science fiction novel, *In the Days of the Comet*.

Perhaps the best place to see the complexity—and the ambiguity—
in Romantic attitudes toward modern science and technology is in
Byron's *Don Juan*. In a somewhat more equivocal tribute to Newton
than Pope's, Byron finds in the legend of Sir Isaac and the apple both
a parallel to man's original fall and the hope of overcoming it:

> When Newton saw an apple, he found
>   In that slight startle from his contemplation—
> 'Tis *said* (for I'll not answer above ground
>   For any sage's creed or calculation)—
> A mode of proving that the earth turned round
>   In a most natural whirl called 'Gravitation,'
> And this is the sole mortal who could grapple,
> Since Adam, with a fall, or with an apple.
>
> Man fell with apples, and with apples rose,
>   If this be true; for we must deem the mode
> In which Sir Isaac Newton could disclose
>   Through the then unpaved stars the turnpike road,
> A thing to counterbalance human woes;
>   For ever since immortal man hath glowed
> With all kinds of mechanics, and full soon
> Steam-engines will conduct him to the Moon.[11]

Though Byron sounds traditional in the way he associates science
with original sin, he also senses the glorious potential in humanity's
newfound technological power and, like Shelley, looks forward to
science as a means of overcoming "human woes." Earlier in the poem,
as Byron contemplates, "What wondrous new machines have late,
been spinning!" (1.130), he views technology as a double-edged
sword:

11. *Don Juan*, Canto x, stanzas 1–2. All quotations from Byron are taken from *The Oxford
Authors: Byron*, ed. Jerome J. McGann (Oxford: Oxford University Press, 1986).

What opposite discoveries we have seen!
  (Signs of true genius, and of empty pockets)
One makes new noses, one a guillotine,
  One breaks your bones, one sets them in their sockets.

                                    (I.129)

For all Byron's skepticism and cynicism, he still sees more hope in
the technological developments of his day than in the political:

This is the patent-age of new inventions
  For killing bodies, and for saving souls,
All propagated with the best intentions;
  Sir Humphrey Davy's lantern, by which coals
Are safely mined for in the mode he mentions,
  Timbuctoo travels, voyages to the Poles,
Are ways to benefit mankind, as true
Perhaps, as shooting them at Waterloo.

                                    (I.132)

Byron was the least idealistic of the English Romantics and certainly
understood how the power of technology could serve evil ends, yet
even he could imagine ways in which it might "benefit mankind."

II

    I have tried to correct the common error that the Romantics were
simply hostile to technology by showing that they rejected only what
they perceived to be its negative effects and in some cases actually
had high hopes that technological developments might help transform the human condition into something paradisiacal. Thus the Romantics criticize modern science and technology only from within
the perspective of modernity. Contrary to common opinion, they do
not reject the modern project as formulated by Bacon; they just want
to make sure that it stays on course to produce a genuinely better
world for humanity. Indeed we must be wary of confusing the essentially modern critique of technology the Romantics offer with the
traditional critique that can be found in the ancients, most powerfully
articulated by Plato and Aristotle. The Romantics may sound like
classical philosophers when they seem to criticize technology in the
name of nature. But we have to be careful not to be misled by words,

for the Romantics do not mean the same thing by the word "nature" that Plato and Aristotle did.

I can indicate the difference by citing one of Blake's Proverbs of Hell: "Where man is not nature is barren."[12] That is a curious statement to be made by a Romantic poet who is supposed to revere nature. It suggests that, for Blake, nature is significant only within a human horizon; it has value only within the context of human purposes. This is the very opposite of the classical understanding of nature, in which man is viewed as problematically trying to fit himself into an overarching natural order greater than he is. To put the point somewhat schematically: Blake judges nature by the standard of man, whereas Aristotle judges man by the standard of nature. This contrast helps to highlight the fact that the nature of Blake and the other Romantics is a thoroughly *humanized* nature, a nature filtered through, represented in, and in the deepest sense produced by a human consciousness. As Percy Shelley writes in his poem "Mont Blanc," "The everlasting universe of things/Flows through the mind" (ll. 1–2). Whether directly or indirectly, the Romantics are all heirs of Kant and his Copernican revolution in epistemology. They may have quarreled with the notion of nature as a construct of pure reason, of Newtonian natural science, but, in part encouraged by Kant's third critique, the *Critique of Judgment*, the Romantics reconceived nature as the construct of the human imagination. In Romantic poetry, nature does not appear as a self-subsistent world, which the poet merely imitates. The Romantics disagreed about the degree of their creativity—in "Tintern Abbey" Wordsworth thinks he can only "half create" the scene he is picturing (l. 106)—but they were agreed that nature cannot exist apart from our representations of it.[13] As deconstructive critics such as Paul de Man have emphasized, we have no immediate access to nature in Romantic literature.[14] Our view of nature is always

12. See *The Marriage of Heaven and Hell*, pl. 10.

13. This controversy was, for example, at the heart of the philosophical differences between Wordsworth and Coleridge. Coleridge's "Dejection: An Ode" tries to rewrite the kind of nature poetry embodied in Wordsworth's "Intimations Ode" by suggesting that the imagination is fed by internal sources, not by the external power of the natural world: "I may not hope from outward forms to win/The passion and the life, whose fountains are within" (ll. 45–46). Coleridge's "To William Wordsworth"—surely one of the most equivocal tributes ever paid by one poet to another—celebrates the achievement of *The Prelude* but argues that Wordsworth has misinterpreted his own creativity: "thy soul received/The light reflected, as a light bestowed" (ll. 18–19). That is, according to Coleridge, Wordsworth interpreted as a gift from nature the power he himself was projecting into nature.

14. See especially Paul de Man, *The Rhetoric of Romanticism* (New York: Columbia University Press, 1984).

mediated by the consciousness of the poet, hence the emphasis on memory and recollection in Wordsworth.

In short, for all the so-called nature piety of Romanticism, in Blake, Wordsworth, and the others nature is nothing without man's art. In view of the fact that the Greek word for art is *technē*, one might venture the paradoxical formulation that for the Romantics nature itself is a technological construct. The Romantics were genuinely concerned about the remaking of nature by the powers of modern technology, but they were at the same time engaged in the project of remaking nature according to their own *technē*. The Romantics do not object to technology on the grounds supplied by classical philosophy: that technology attempts to go beyond the natural limits of humanity, to transgress the boundaries nature sets to man's desires and aspirations. On the contrary, the Romantic attitude toward technology reflects precisely a sense of the unlimited character of human power, rooted in the infinity of humanity's desires and imagination.

The scholars who reject the notion that the "Satanic Mills" in *Milton* refer to industrialization often trace the term instead to this passage in Blake's early treatise *There Is No Natural Religion:*[15]

> IV   The bounded is loathed by its possessor. The same dull round even of a universe would soon become a mill with complicated wheels.
>
> V   If the many become the same as the few, when possess'd, More! More! is the cry of a mistaken soul, less than All cannot satisfy Man.
>
> VI   If any could desire what he is incapable of possessing, despair must be his eternal lot.
>
> VII   The desire of Man being Infinite the possession is Infinite & himself Infinite.

I know of no passage that better encapsulates the spirit of Romanticism. Blake presents the natural world conceived by modern science as a construct that artificially distorts and restricts human perception and desire. He wants to break out of this picture of the universe, with its clockwork wheels within wheels, not to return to the classical understanding of nature as a self-subsisting order but to usher in an imaginative recreation of the world in which all human desires can

---

15. See, for example, the note on the passage in *The Norton Anthology of English Literature,* 5th ed., ed. M. H. Abrams (New York: Norton, 1986), 2:77.

be liberated and fulfilled. The Romantic critique of technology is not rooted in the classical concern that it inflames human desire but in a radically modern objection that it, in fact, restrains desire. Strictly speaking, the Romantics object to technology, not because it is unnatural but because it is unimaginative.[16] In fact, there is nothing of a genuine piety toward nature in Romanticism. The fundamental Romantic assumption is that the human spirit is the greatest power in the universe, to which anything we understand as nature is profoundly subordinate. Wordsworth's *Prelude* builds up to a vision in which "the mind of Man becomes/A thousand times more beautiful than the earth/On which he dwells" (xiv.450–52).

If we did not know that this is the great nature poet Wordsworth speaking, we might be tempted to characterize this passage as impiety to nature. And in the most imaginative treatment of the problem of technology in the Romantic era, the whole Romantic project is presented as unnatural and impious, a satanic rebellion against the limits of the human condition. I am speaking of Mary Shelley's *Frankenstein*, generally considered the first authentic example of science fiction. Those who think of science and literature as simply opposed in Romanticism would have a hard time accounting for this novel because Mary Shelley assimilates her image of the scientist to her image of the artist. For example, Victor Frankenstein speaks of his creature as "a thing such as even Dante could not have conceived"[17]; in discussing her hero in the preface, Mary Shelley refers to how "his success would terrify the artist" (p. xi). Moreover, as many critics have pointed out, she used several details from the life of her husband, Percy, in creating the portrait of Frankenstein (for example,

16. Consider, for example, the following passage from Shelley's *Defence:* "We want the creative faculty to imagine that which we know; we want the generous impulse to act that which we imagine; we want the poetry of life. . . . The cultivation of those sciences which have enlarged the empire of man over the external world, has, for want of the poetical faculty proportionally circumscribed those of the internal world; and man, having enslaved the elements, remains himself a slave. To what but a cultivation of the mechanical arts in a degree disproportioned to the presence of the creative faculty . . . is to be attributed the abuse of all invention for abridging and combining labour to the exasperation of the inequality of mankind? From what other cause has it arisen that the discoveries which should have lightened, have added a weight to the curse imposed on Adam?" (502–3). This passage is crucial for understanding the peculiar Romantic attitude toward technology. We see here that Shelley accepts the Baconian project of conquering nature and merely wants to avoid its abuses. In particular, he hopes to rely on technology for labor-saving devices to reduce human misery. Shelley proposes a technology guided by poetry, which will supply it with genuinely imaginative goals.

17. Mary Shelley, *Frankenstein* (New York: New American Library, 1965), p. 57.

Victor was Percy's childhood name for himself).[18] Far from seeing
modern science and Romanticism as opposed, Mary Shelley suggests
a profound connection between the two phenomena. Frankenstein
attempts in physical terms what Percy Shelley and the other Romantics were attempting to do in spiritual or poetic terms: to make a
better human being. The novel imaginatively recreates the world of
late eighteenth-century Europe, with all its revolutionary fervor and
apocalyptic expectations for a radical transformation of the human
condition. Frankenstein may work in flesh and blood, but he is at
heart a Romantic poet, with an ideal vision of how man might be
made perfect, of how he as creator could improve upon the handi-
work of nature and/or God.

Unfortunately, like a Romantic poet, Frankenstein finds that when
he attempts to embody his ideal vision in concrete form, it fails to
live up to his great expectations: "His limbs were in proportion, and
I had selected his features as beautiful. Beautiful! Great God! His
yellow skin scarcely covered the work of muscles and arteries beneath" (p. 56). In many ways *Frankenstein* retells the tragedy of the
French Revolution, chronicling how initially idealistic impulses, when
put into practice, get twisted and distorted into a horrible parody of
the original hopeful vision. Having dreamed of outdoing nature as a
creator, Frankenstein finds that he has produced something monstrous. Mary Shelley acutely sees through the smokescreen of nature
piety in Romanticism. She has Frankenstein speak with awe of natural
scenery, but he has no reverence whatsoever for human nature. On
the contrary, he is so unimpressed by it that he wants to tinker with
it. He refuses to accept the natural limits of humanity; in particular,
he wishes to overcome human mortality. The author shows her idealistic creator willfully appropriating the traditional prerogatives of God
for himself. Overflowing with the rhetoric of the Enlightenment,
Frankenstein reveals how much he is motivated by an almost Nietzschean will to power: "Life and death appeared to me ideal bounds,
which I should first break through, and pour a torrent of light into
our dark world. A new species would bless me as its creator and
source. . . . No father could claim the gratitude of his child so completely as I should deserve theirs" (p. 52). With remarkable pre-

18. See Christopher Small, *Mary Shelley's Frankenstein: Tracing the Myth* (Pittsburgh: University of Pittsburgh Press, 1973), p. 101. Chap. 4 of *Creature and Creator* is devoted to *Frankenstein;* there I develop more fully the interpretation outlined here; the notes to that chapter
indicate the critical works that have influenced my thinking about Mary Shelley.

science, Mary Shelley uncovers the tyrannical impulses that often lurk beneath projects for improving humanity, whether proposed by modern scientists or Romantic artists.

What is particularly revealing in this passage is Frankenstein's assertion: "No father could claim the gratitude of his child so completely as I should deserve theirs." Frankenstein's experiment is an elaborate detour around normal biological procreation. He wants to have a child without having to share credit with a partner, the mother. No doubt Mary Shelley, as a woman, was sensitive to the irony of Frankenstein's situation; he keeps pushing aside his fiancée, Elizabeth, in a monomaniacal attempt to create life all by himself, thereby neglecting the simpler, customary, and more enjoyable way.[19] But Frankenstein is a true Romantic—everything must come out of himself. He will do anything to avoid being indebted to nature. In her portrait of Frankenstein, the Romantic visionary, Mary Shelley shows how the spirit of modernity cuts across the seeming boundary between science and literature, and the unifying principle is the technological character of modernity, broadly conceived. Modernity is the impulse to transform nature in accordance with human desire, culminating in the effort to transform human nature itself. In its attitude toward human nature, Romanticism ultimately proves to be no more inclined than modern technology to respect natural limits.

III

Thus in *Frankenstein* Mary Shelley provides an insightful critique of the affinities between Romanticism and the technological impulse. The way she portrays Frankenstein going about the actual business of creation helps us to gain a fuller understanding of what is distinctive about the Romantic attitude toward nature. Despite his hopes, Frankenstein cannot dispense with all aid from nature. He cannot build his creature from scratch; he must turn to nature's junkyard to get the spare parts: "I collected bones from charnel-houses. . . . The dissecting room and the slaughter-house furnished many of my materials" (p. 53). It is curious that Frankenstein never thinks of animating a corpse, even though bringing dead people back to life seems to be

19. See Robert Kiely, *The Romantic Novel in England* (Cambridge: Harvard University Press, 1972), p. 164.

one of the principal motives of his experiments. But if he animated a corpse, Frankenstein could claim to be only its re-creator, not its creator. Only if he assembles body parts into a new and never before existing whole can he still think of himself as genuinely creative in the Romantic sense. Though forced to rely on nature for parts, Frankenstein keeps his indebtedness to a minimum by not attempting to duplicate any natural order of the parts, hence his unwillingness to work with whole bodies. He maintains his integrity as a creator precisely by not respecting the integrity of nature.

This attitude toward nature is characteristic of Romanticism in general. The Romantics are preoccupied with natural scenery, but they tend not to attempt to image nature as a whole. That is, the Romantics do not think of nature as a cosmos, as a natural *order,* as a whole articulated into intelligible forms. Rather they tend to be attracted to scenes of nature in disorder, scenes of great destruction like earthquakes or avalanches, scenes that suggest a chaos rather than a cosmos in nature. Percy Shelley's "Mont Blanc" is a case in point: the poem implies that we can never grasp the natural phenomenon as a whole. If nothing else, the alpine peak remains obscured, like a Kantian *noumenon:* as human observers we have no access to the essence of the thing-in-itself. We are left with a chaotic jumble of phenomena that elude our attempts to categorize them:

> a flood of ruin
> Is there, that from the boundaries of the sky
> Rolls its perpetual stream; vast pines are strewing
> Its destined path, or in the mangled soil
> Branchless and shattered stand; the rocks, drawn down
> From yon remotest waste, have overthrown
> The limits of the dead and living world
> Never to be reclaimed. (ll. 107–14)

This view of a nature existing only in fragments reappears in a similar alpine scene in *Frankenstein:*

> The abrupt sides of vast mountains were before me; the icy wall of the glacier overhung me; a few shattered pines were scattered around; and the solemn silence of this glorious presence-chamber of imperial nature was broken only by the brawling waves or the fall of some vast fragment, the thunder sound of the avalanche or the cracking, reverberated along the mountains, of the accumulated

ice, which, through the silent working of immutable laws, was ever
and anon rent and torn, as if it had been but a plaything in their
hands. (p. 92)

Here we see clearly the peculiar inflection of the new Romantic view
of nature. What strikes the Romantic is not the seamlessness of nature's joints, as it were, but the abruptness of its cracks, not how
nature holds together but how it falls apart. If the world is not a
natural whole, then it can be made into a whole only by human
effort, which places the Romantic imagination just where it wants to
be: at the center of the universe.

Mary Shelley's mention of the "silent working of immutable laws"
points to the origin of this change in the Romantic view of nature.
Frankenstein portrays a Newtonian world in which nature operates
blindly according to necessity.[20] As paradoxical as it may sound, the
Romantics accepted the reconception of nature brought about by
modern science. In the classical understanding the universe is ordered in terms of natural ends. In the traditional Christian view the
world is ordered by divine providence. In modern science nature
becomes fundamentally a chaos, a world of matter in random motion,
obeying observable laws, but laws that increasingly have come to be
understood as merely statistical in character.[21] It seems odd that the
Romantics should have accepted such a bleak and soulless view of
the universe, but this kind of cosmology was, in fact, necessary to
provide support for their view of human freedom. A universe of
indeterminacy allows for indeterminacy in human action. The Romantics actually want the universe to be soulless in itself so that they
can be the ones to inject soul into it. If one wants soul, as it were,
built into the universe, one must accept the ancients' teleological understanding of nature; but if one grants that nature is shot through
with purpose, one must accept the idea that human beings must take
their bearings in their actions from their natural end. The Christian
view of the universe is even more teleological. Once nature is satu-
rated with divine purpose, the freedom of human beings to create
their own ends is severely restricted. As strange as it sounds, the
Romantics welcomed the antiteleological view of the universe devel-

20. On the subject of necessity in Romantic thought, see the long note Percy Shelley
wrote on the issue for his early poem *Queen Mab*. The note is available in *Shelley's Prose*, ed.
David Lee Clark (Albuquerque: University of New Mexico Press, 1954), pp. 109–12.

21. The recent development of the science of chaos confirms this notion. See James Gleick,
Chaos: Making a New Science (New York: Viking, 1987).

oped in modern science because it liberated them from having to submit to the standard of natural ends, derived from a fixed conception of human nature. Though many intellectual historians try to contrast the organic view of nature in Romanticism with the mechanistic view in the Enlightenment,[22] Romantic organicism did not displace the mechanistic world view of Newtonian science but was instead built upon it. Simply stated, the Romantics were not Aristotelians; they did not return to a teleological view of nature. The great magic trick of Kant's *Critique of Judgment* was to try to show how the concept of the organic could be made compatible with a mechanistic view of nature. Kant's juggling with the idea of "purposefulness without purpose" (*Zweckmässigkeit ohne Zweck*) was a desperate attempt to save organic phenomena without having to return to a fully developed notion of teleology in nature.[23]

This consideration explains Kant's importance to the Romantics. If nature is not orderly in the classical or Christian sense, Kant's solution becomes particularly attractive to the Romantics. Whatever order nature has is supplied by human reason, or, in the Romantic revision of Kant, human imagination. Kant's reconception of the idea of the sublime pointed in this direction. He rethought sublimity so that it ceases to be a quality inhering in natural phenomena and becomes instead an aspect of the mind contemplating nature. For Kant the disordered, violent, and monstrous in nature all become reinterpreted as evidence of the sublime power of the human spirit. We prove our power to transcend nature by our unwillingness to submit to the seemingly overpowering forces represented by these violent phenomena. Just when nature threatens to master us, we display the power of our imagination to master nature.

That then is the fundamental point of contact between Romanticism and modern technology: they share the impulse to master nature. The Romantic love of nature is, in effect, a form of *agape*, not *eros*. That is, the Romantics do not look up to nature as something higher than humanity, a larger order in which human beings might seek and perhaps even find a way to complete themselves. Rather

22. For an example of this kind of argument, see Hans Eichner, "The Rise of Modern Science and the Genesis of Romanticism," *PMLA* 97 (1982), 8–30. In general, this essay is an excellent example of the kind of understanding of the relation of Romanticism to modern science that I am arguing against here. The classic statement of this view is the chapter "The Romantic Reaction," in Whitehead, *Science and the Modern World*, pp. 75–94.

23. I have discussed this matter at some length in my essay "The Metaphysics of Botany: Rousseau and the New Criticism of Plants," *Southwest Review* 70 (1985), 362–80.

the Romantics look down upon nature as something perhaps beautiful but nevertheless subject to humanity's creative power and capable
of being improved upon.[24] As *Frankenstein* suggests, the Romantics
try to appropriate for themselves the traditional position of the Christian God vis-à-vis the natural world. There is often something almost
patronizing about the Romantics' attitude toward nature. When they
are not picturing nature in violent upheaval, they tend to concentrate
on fragile flora and fauna, sometimes, as in Coleridge's "Rime of the
Ancient Mariner," with an explicit message of being kind to lower
creatures. For all their admiration of nature, and even Wordsworth's
tendency to speak of it as a moral teacher, the Romantics on the
whole tend to regard the natural world as a lower order of being,
which because of the creative power of human beings, technological
and otherwise, has passed under their supervision and protection,
which is to say, their control.

IV

This analysis of the Romantic attitude toward technology may seem
at first a purely antiquarian enterprise, but I hope that it has made
clearer that in many ways we today have not gotten beyond the Romantics in our thinking about the problem. I do not have the space
in this essay to develop the parallels at length but conclude with a
few basic observations. One lesson we can learn from the Romantics
is the danger of remaining locked within a technological perspective
even when we think we are criticizing technology. So great are its
material benefits that we moderns have difficulty contemplating the
genuine rejection of technology even when we become aware of its
pitfalls. The great temptation is to think that the problem is merely
one of modifying our use of technology, purging it of its evils while
retaining all of its benefits. That is why it is rare to find a truly radical
critique of technology in the modern world, though it has been possible at least since Heidegger.

Thus thinking about the Romantic view of technology can help
clarify our own understanding of the subject, particularly when we
reflect on how the meaning of *nature* became transformed in Romantic
discourse. We can observe the logical outcome of this process in the

24. On this point, see especially Shelley, *Prometheus Unbound*, ii.iii.11–16.

fact that today when people voice their concern for nature, they often choose to voice their concern for the environment.[25] What does this semantical change mean? One can be in awe of nature, but one cannot be in awe of the environment. In a way, this change of terminology perfectly embodies the transformation Romanticism helped bring about in our conception of the world around us, namely, reconceiving the world precisely as centering around *us*. The environment is, in fact, nature reconceived as lying purely within a human horizon. The environment is *our* world, *our* home, supposedly to protect and cherish but basically to do with as we wish. We talk so much these days about preserving nature only because we have become convinced that we now have it within our power to destroy it. Calls for the protection of the environment, however noble their rhetoric, are often couched in the most utilitarian terms. We need to preserve the tropical rain forests because our breathable air may be at stake. That is, nature must be protected, not because of any intrinsic merit it may have but because it is instrumental to human good. There is very little underlying respect or reverence for nature in such arguments, only a concern that if we annihilate nature we may annihilate ourselves. Talk of ecosystems has replaced talk of the natural order. The reason is that nobody ever felt constrained to obey an ecosystem, whereas the idea of a natural order used to convey significant ethical implications.

What we saw in *Frankenstein* helps to suggest another problem in the way we now think about the relation between technology and nature. Like Victor Frankenstein, we have deflected whatever reverence we have left away from ourselves onto nonhuman nature; indeed today we tend to have regard for nature in every species but our own. It is indeed a curious spectacle to see, as often happens, that the very same people who are deeply concerned about humans tampering with some species like the manatee are also welcoming as progress all sorts of technological interference in the biological functioning of the human species, especially in the area of reproduction. Somehow we can be Aristotelians when it comes to snail darters, insisting upon the sanctity of species in the case of lower creatures, but Darwinians when it comes to our own species, rushing to help nature along on the path of evolution. This tension in our thinking can partly be traced to the influence of Romanticism, which, as I have tried to show,

25. For this observation, I am indebted to Michael Valdez Moses.

in effect redirected piety from the higher to the lower. The Romantic's heart goes out to all the beings beneath him, in part because he can no longer imagine a being above him. Under these circumstances he is happy to preserve the status quo for all the lower species but will not brook limits on what he thinks of as his own progress.

Since by speaking of *Frankenstein* I have introduced the subject of science fiction, allow me to conclude by referring to one of the more profound explorations of the problem of technology in our own day, *Star Trek*. If one surveys the voyages of the Starship *Enterprise,* in both television and motion pictures, one will note a curious pattern in Captain Kirk's cumulative achievement: he managed to save the humpbacked whale in *Star Trek IV,* but in a television episode he destroyed the power of the Greek god Apollo (an episode, incidentally, that bore the exceedingly Romantic title "Who Mourns for Adonais?"). "Save the whales; kill the gods": I do not believe that these actions are unrelated; rather I find this conjunction strangely emblematic of the whole interplay of Romanticism and technology in the modern world. The basis of our solicitude for all the beings beneath us in the universe has been the eradication of any sense of beings above us. Somehow the producers of *Star Trek* intuited that this process required specifically the destruction of something Greek. In another television episode, Kirk also gets to undo a community based on Plato's *Republic.* Indeed the famous five-year mission of the *Enterprise* to seek out new life and new civilizations mysteriously metamorphosed week after week into a search-and-destroy mission. Though not even Mr. Spock could formulate the principle, a study of *Star Trek* in detail would show that the real mission of Kirk and his crew was to stamp out any form of life in the universe that might possibly lay claim to being higher than humanity, especially any lingering form of aristocracy in the universe. As the wake of destruction the *Enterprise* left behind in the galaxy testifies, when man chooses to revere nothing higher than himself, he will indeed find it difficult to control the power of his own technology.

Suggestions for Further Reading

Chapple, J. A. V. *Science and Literature in the Nineteenth Century.* London: Macmillan, 1986.
Cunningham, Andrew, and Nicholas Jardine, eds. *Romanticism and the Sciences.* Cambridge: Cambridge University Press, 1990.

Levine, George, and U. C. Knoepflmacher, eds. *The Endurance of Frankenstein.*
Berkeley: University of California Press, 1979.

Ralston, Gilbert A., and Gene L. Coon, "Who Mourns for Adonais?" In *Star Trek*
7, ed. James Blish. New York: Bantam Books, 1972.

PART II

The Destiny of Modern Technology

6 Self-Ownership, Communism, and Equality: Against the Marxist Technological Fix

G. A. COHEN

I

In this essay I argue that Marxist reliance on material superabundance as the solution to social problems is connected with Marxist reluctance to effect an absolute break with certain radical bourgeois values.[1] The "Marxist technological fix" has served as a means of avoiding questions about justice which those who seek to carry the Marxist tradition forward cannot now conscionably ignore.

I shall call the bourgeois thought structure from which, so I claim, Marxism has failed to distinguish itself (sufficiently thoroughly) "left-wing libertarianism." Because the meaning I assign to that phrase is not the only one it could reasonably be thought to bear, I must explain my use of it here.[2]

---

1. The first seven sections of this essay were previously published, in a slightly different form, in *Proceedings of the Aristotelian Society, Suppl. Vol.*, 1990, and are reprinted courtesy of the editor of the Aristotelian Society, © 1990.

2. The first two sections of this essay revise and expand pp. 114–17 of my "Self-Owner-ship, World-Ownership, and Equality," which appeared in *Justice and Equality Here and Now*, ed. Frank Lucash (Ithaca: Cornell University Press, 1986). In that paper, what is here called "libertarianism" was called "liberalism" because of *one* of the senses that the latter term has traditionally borne, but I have been persuaded that it makes for confusion so to use "liberal" that John Rawls and Ronald Dworkin, who are widely and rightly called liberals, come out emphatically anti-liberal. Hillel Steiner uses "liberal" in the old and disappearing sense I

131

A libertarian, in the present sense, is one who affirms the principle of self-ownership, which occupies a prominent place in the ideology of capitalism. The principle says that every person is morally entitled to full private property in her own person and powers. This means that each person has an extensive set of moral rights (which the law of her land may or may not recognize) over the use and fruits of her body and capacities, comparable in content to the rights enjoyed by one who has unrestricted private ownership of a piece of physical property.

One right in physical property that is especially important for the explication, by analogy, of the content of self-ownership is the right not to be forced to place what one owns at the disposal of anyone else. My land is not *fully* mine if someone else has a right of way over it or a claim to a portion of the income it generates. Analogously, I am not fully mine, I do not fully own myself, if I am required, on pain of coercive penalty, and without my having contracted to do so, to lend my assistance to anyone else, or to transfer (part of) what I produce to anyone else, either directly or through state-imposed redistribution. This prohibition on forcing a person to bestow her service or product on another follows from the freedom to use her powers as she wishes with which each person is endowed under the self-ownership principle. Whatever else self-ownership means, and its meaning is, within limits, debatable,[3] the stated prohibition is one of its entailments, and, moreover, the polemically crucial one, the entailment of chief significance for controversy in political philosophy.[4]

The libertarian principle of self-ownership has been put to both progressive and reactionary use, in different historical periods. It was put to progressive use when it served as a weapon against the noncontractual claims of feudal lords to the labor of their serfs. By contrast, it is, in our own time, put to reactionary use, by those who argue that the welfare state unjustifiably enforces assistance to the needy. In this essay I largely abstract from differences in the import of the principle that reflect differences in the historical context of its application. What interests me here are differences in its import that

had in mind in his "Liberal Theory of Exploitation" (*Ethics* 94 [January 1984], 225–42), and Antony Flew laments the passing of that sense in his entry on "libertarianism" at p. 188 of his *Dictionary of Philosophy* (London, Fontana, 1979).

3. See the Appendix at the end of this essay.

4. I offer a further defense of the claim that self-ownership possesses the stated entailment in a discussion of David Gauthier in a forthcoming book on equality.

emerge when libertarianism is conjoined with other principles, as I shall now explain.

Libertarianism, as I have defined it, may be combined with contrasting principles with respect to those productive resources that do not inhere in persons, to wit, the substances and powers of nature. As a result, libertarianism comes in both right- and left-wing versions. All libertarians say that each person has a fundamental entitlement to full private property in herself and, consequently, no fundamental entitlement to private property in anyone else. (The qualification "fundamental" caters for the fact that, just as I may transfer my physical property to another, so I may transfer my ownership of myself and become another's slave: full self-ownership allows—indeed, it legitimates—derived rights of property in others and a corresponding derived lack of property in oneself.) Right-wing libertarianism, of which Robert Nozick is (or was) an exponent,[5] adds[6] that self-owning persons can acquire similarly unlimited original rights in unequal amounts of external natural resources. Left-wing libertarianism is, by contrast, egalitarian with respect to initial shares in external resources: Henry George, Leon Walras, Herbert Spencer, and Hillel Steiner have occupied this position.[7]

Two major contemporary thinkers whose doctrines embody a rejection of libertarianism are Ronald Dworkin and John Rawls. They would, to a certain significant extent, restrict people's self-ownership,[8] for they say that people's talents do not, in every respect, morally speaking, belong to them, but may, for certain purposes, be regarded

5. In my view, the idea of self-ownership (and not the concept of freedom: see my "Self-Ownership, World-Ownership, and Equality: Part II," in *Marxism and Liberalism*, ed. Ellen Paul et al. [Oxford: Blackwell, 1986], p. 77, n. 2) is the most likely point of departure for generating the rights that Robert Nozick has said we all have, but which he has never explicitly specified. For invocations of self-ownership (and disparagement of its contrast, the ownership of a person by others) by Nozick, see *Anarchy, State, and Utopia* (New York: Basic Books, 1974), pp. 172, 281–83, 286, 290.

6. Some right-wing libertarians would regard what follows not as an addition to but as an entailment of the principle of self-ownership; see note 7.

7. I am here bracketing the question of the relationship between the self-ownership tenet and the tenet about resources in each variant of libertarianism. Some might hold that self-ownership requires left-wing libertarianism, others that it requires right-wing libertarianism, and still others that the question of justice with respect to external resources is not settled by the self-ownership thesis. (In "Self-Ownership, World-Ownership, and Equality" I criticized what might be described as Nozick's attempt to derive right-wing libertarianism from libertarianism.)

8. Or so I say: I do not claim that Dworkin and Rawls would find that description of their views appealing.

as resources over the fruits of whose exercise the community as a whole may legitimately dispose.

Now, in my contention, Marxists must oppose what I have called left-wing libertarianism more forthrightly than they have done if they are to be true to some of their most cherished beliefs. I do not here claim that Marxists have accepted the libertarian principle,[9] but that, in their handling of two large issues, they have failed frontally to reject it. I believe that those failures to reject disfigure the standard Marxian treatment of the two issues, and that, for superior treatment of them, in the spirit of Marxism, it is necessary to embrace the anti-self-ownership tenet attributed to Rawls and Dworkin above.

(Note that, in urging Marxists to reject the principle of self-ownership, I do not say that they should affirm no self-ownership rights at all. Rejecting the principle of self-ownership is consistent with affirming some self-ownership rights, and I shall discuss elsewhere which self-ownership rights, if any, Marxists should affirm.)

II

The first area in which Marxists have failed to oppose left-wing libertarianism is in their critique of capitalist injustice. In the Marxian version of that critique, the exploitation of workers by capitalists, that is, capitalists' appropriation without return of part of what workers produce, derives entirely from the fact that workers lack access to physical productive resources and must therefore sell their labor power to capitalists, who enjoy a class monopoly in those resources. Hence, for Marxists, capitalist appropriation is rooted in an unfair distribution of rights in external things. The appropriation has its causal origin in an unequal distribution of productive resources, and it suffices for considering it unjust exploitation that it springs from that initial unjust inequality. The Marxist critique of capitalist appropriation thus requires no denial of the thesis of self-ownership.[10]

---

9. That claim might, nevertheless, also be mounted on the basis of certain Marxist understandings of the grounds for saying that capitalists exploit workers: see my "Marxism and Contemporary Political Philosophy, or: Why Nozick Exercises Some Marxists More than He Does Any Egalitarian Liberals," *Canadian Journal of Philosophy, Suppl. Vol.* 16 (1990).

10. In chap. 24 of vol. 1 of *Capital* (Harmondsworth, Middlesex: Penguin Books, 1976), Marx purports to show that exploitation can also occur without an original unfair distribution of external resources. He grants, for the sake of argument, that the capitalist's initial capital was accumulated "from his own labour and that of his forefathers" (p. 728). Once possessed of that capital, however, he causes it to grow through the appropriation of unpaid labor, and

When social democrats (or liberals, in the American sense of the term) call for state intervention on behalf of the less well off, they are demanding that the better off lend them assistance, and they are, accordingly, rejecting the thesis of self-ownership. The social democratic attitude was visible in those anti-Thatcherite *Guardian* leader columns that complained about the fate of those who are ill equipped to provide for themselves in market competition, however "fair" it may be. The political rhetoric of Marxists is quite different. They do not, in their critique of capitalist injustice, demand that the well off *assist* the badly off. In the Marxist focus, the badly off people under capitalism are the proletariat, and they are badly off because well-off people, or their forebears, have dispossessed them. What afflicts the badly off is that they are forcibly denied control of physical resources, and, under that construal of the plight, the demand for its redress needs nothing stronger than left-wing libertarianism as its ground. In the Marxist claim, the badly off suffer injustice in the left libertarian sense that they do not get their fair share of the external world.

Now, the external resources usually mentioned by Marxists in the present connection are means of production. Marxists say that what makes possible the exploitation of workers by capitalists is that workers lack means of production, which lie in the capitalists' hands. Some means of production are not, however, natural resources but products of labor, such as tools, machines, and computers, and, since I characterized left-wing libertarians as egalitarian with respect to natural resources only, it might seem that, in their protest against workers' lack of means of production, Marxists are, contrary to my contention, relying on something different from the left-wing libertarian norm. But that is not so. For those means of production that are not natural resources are the product, in the end, of natural resources and labor power, "the two primary creators of wealth,"[11] and since workers under capitalism do not lack ownership of their own labor power,[12]

---

hence through unjust exploitation, until "the total capital originally advanced becomes a vanishing quantity" (p. 734). I argue elsewhere that this conception of exploitation both presupposes *and* rejects the principle of self-ownership: see the discussion of "clearly generated capitalism" in sec. 8 of "Marxism and Contemporary Political Philosophy." I cannot go into that complex paradox here.

11. Ibid., p. 752. Marx identifies those "creators" as "labour-power and land" (or, p. 752, as "man and nature"), and he explains (p. 758) that "land" "means, economically speaking, all the objects of labour furnished by nature without human intervention."

12. It suffices here that they enjoy that ownership as a matter of formal legal fact, even if, because they own no means of production, they are *in effect* as lacking in self-ownership as serfs are: see "Marxism and Contemporary Political Philosophy," sec. 2.

it must, ultimately, in the Marxist view, be their lack of ownership of natural resources that accounts for their vulnerability to exploitation by capitalists.[13]

Now, Marxist nonopposition to left libertarianism in the matter of the critique of capitalist injustice cannot be sustained. What Marxists regard as exploitation, to wit, the appropriation, without return, of surplus product, will indeed result when people are forcibly denied the external means of producing their existence. One case of that is what Marx called "primitive accumulation," the process whereby, in his account of it, a relatively independent British peasantry was turned into a proletariat by being deprived of its land. But such forced dispossession, while assuredly a sufficient causal condition of what Marxists think is exploitation, is not also a necessary condition of it. For if all means of production were distributed equally across the population, and people retained self-ownership, then differences in talent and time preference and degrees of willingness to take risk would bring about differential prosperity that would, in due course, enable some to hire others on terms that Marxists would regard as exploitative—at least at levels of development of the productive forces below those at which, according to Marxists, capitalism, and there-fore capitalist exploitation, will not obtain. That is, with very ad-vanced productive forces, which enable even relatively unproductive people to produce a lot without assistance, equal distribution of exter-nal resources might not lead to exploitation, but you do not, in any case, on a Marxist view, get *very* advanced productive forces while capitalism still prevails. (It is, of course, difficult to imagine each of n individuals being privately endowed with one-nth of the productive forces of a modern society, but essentially the same point applies if we begin with a more feasible hypothesis: if all means of production were socially owned and leased [renewably] to workers' cooperatives for finite periods, then, once again, differences other than those in initial resource endowments could lead to indefinitely large degrees of inequality of position and, from there, to exploitation, with some co-ops in effect exploiting others.)

Marxists have, then, exaggerated the extent to which what they consider exploitation depends on an initial inequality of rights in worldly assets. The story about the dispossession of the peasants from the soil does not impugn capitalism as such. It impugns only

13. For further development of this claim, see my "Marx and Locke on Land and Labour," *Proceedings of the British Academy* 71, 1985 (Oxford: Oxford University Press, 1986), sec. 2.

capitalisms with one sort of (dirty) prehistory. Even left-wing libertarians would condemn capitalisms with that sort of prehistory, but Marxists are against capitalism as such and must therefore condemn a capitalism in which the exploitation of workers comes from the exploiters' deft use of their self-owned powers on the basis of no special advantage in external resources. Libertarians could not call that exploitative, but Marxists must, and so, to prevent what they consider exploitation, equality with respect to external resources is not enough, yet they proceed in their critique of capitalism as though it would be. (I do not mean that Marxists think that external-resource equality would eliminate exploitation. They amply realize that it would not. But that is inconsistent with their claim that lack of means of production is the root cause of exploitation.) To block generation of the exploitation characteristic of capitalism, people must have claims on the fruits of the powers of other people, claims denied by left-wing libertarianism.

I should add that left-wing libertarians are themselves frequently unaware of how unstable the initial equality they favor would be. Their attempt to combine the attractions of egalitarianism and self-ownership is bound to fail, and the implications of their view (as opposed to its motivation) are much closer to those of right-wing libertarianism than they readily realize. (The work of Hillel Steiner illustrates this claim.)[14]

III

The second important topic, in their reflection on which Marxists fail frontally to oppose left-wing libertarianism, is the nature of the good society. I do not mean that all left-wing libertarians have the same vision of the good society as Marxists do, but rather that no one could base any normative objection to Marx's vision of it on left-wing libertarianism, considered just as such. For in Marx's good society, productive resources are communally owned (as they are in extreme left libertarianism), but the individual remains effectively sovereign over himself. He conducts himself "just as he has a mind,"[15]

14. For a pertinent discussion of Steiner's work, which offers something similar to the "bound to fail" verdict that I venture above, see Eric Mack, "Distributive Justice and the Tensions of Lockeanism," in Social Philosophy and Policy 1 (Autumn 1983).

15. Karl Marx and Frederick Engels, The German Ideology (1846), in Marx/Engels: Collected Works (London: Lawrence and Wishart, 1975), 5:47.

developing himself freely not only without blocking the free development of others but even as a condition of the free development of others. A circumstance of overflowing abundance renders it unnecessary to *press* the talent of the naturally better endowed into the service of the poorly endowed for the sake of establishing equality of condition.

One way of picturing life under communism, as Marx conceived it, is to imagine a jazz band in which each player seeks fulfillment as a musician. Though basically interested in personal fulfillment, and not in that of the band as a whole, or of her fellow musicians taken severally, she nevertheless is maximally fulfilled only to the extent that each of the others also is, and the same holds for each of them. There are, additionally, some less talented people around who obtain high satisfaction not from playing but from listening, and their presence further enhances the fulfillment of the band's members. Against a backdrop of abundance, players and audience alike pursue their own several bents: no one cleans the streets (unless she enjoys doing that) or spends her life as an appendage to a machine. Everyone is guided by her self-regarding goal, yet there is no inequality in the picture to exercise an egalitarian.

So, as I understand Marx's communism, it is a concert of mutually supporting self-fulfillments in which no one considers promoting the fulfillment of others any kind of obligation. I am not, of course, denying that each delights in the fulfillment of the others. Unless they are crabby people, they probably do so. But no such delight is demanded by the *Manifesto* slogan. It is not something in the dimension of affect that is supposed to make communism possible. Instead, a lofty material endowment ensures that "the free development of each is the condition for the free development of all."[16]

In speaking of the Marxist vision of the good society, I have in mind Marx's "higher phase of communist society,"[17] which later Marxists came to call communism. Marx made a distinction between lower

16. Marx and Engels, *Manifesto of the Communist Party* (1848), in *Collected Works*, 6:506. Compare *The German Ideology*, p. 78: "Only within the community has each individual the means of cultivating his gifts in all directions." Note that community is here a *means* to the independently specified goal of the development of each person's powers. See further, *Capital*, 1:618, where Marx insists that "the partially developed individual, who is merely the bearer of one specialized social function, must be replaced by the totally developed individual, for whom the different social functions are different modes of activity he takes up in turn."

17. Karl Marx, *Critique of the Gotha Programme*, in *Marx/Engels: Selected Works in One Volume* (London: Lawrence and Wishart, 1968), p. 324.

and higher phases of communist society, and subsequent tradition renamed it by distinguishing between socialism and communism. What the subsequent tradition has called socialism—Marx's lower phase—is, for Marx, a society in which means of production are owned by the community as a whole and in which, after certain deductions from it have been made, people receive from the total social product something *"proportional* to the labour they supply,"[18] so that those who contribute more get more: I shall call that the "socialist proportionality principle." (The deductions effected before the principle is applied cover depreciation and investment, insurance and contingency funds, the costs of government, the supply of schooling, hospitals, and other facilities catering to "the common satisfaction of needs," and, finally, funds for those who are unable to work.)[19]

Marx thought that socialism would be a great advance over capitalism, under which some able-bodied people live without laboring at all by appropriating the fruits of the labor of others. But he also thought that it would be only a restricted advance. No able-bodied person could live without working because the basis for doing so, private ownership of the means of production, would have disappeared; but Marx believed that the socialist proportionality principle had its own limitations, and I shall now examine some of his criticism of it.

"The right of the producers," Marx says, "is proportional to the labour they supply," and he regards that condition as an advance on capitalist society. But "in spite of this advance, this equal right is still constantly stigmatized by a bourgeois limitation":

> [For] one person is superior to another physically or mentally[20] and so supplies more labour in the same time, or can labour for a longer time. . . . This equal right is [therefore] an unequal right for unequal labour. It recognizes no class differences, because everyone is only a worker like everyone else; *but it tacitly recognizes unequal individual endowment and thus productive capacity as natural privileges.*[21]

18. Ibid., p. 324.
19. Ibid., pp. 322–23.
20. For an inexplicable failure to register the import of the foregoing part of Marx's sentence and a consequent entirely misplaced criticism of Marx's claim that socialism violates the idea of equality, see Richard Norman, *Free and Equal* (Oxford: Oxford University Press, 1987), p. 115.
21. *Critique of the Gotha Programme,* p. 324 (emphasis added). The further passages from the *Critique* quoted below follow more or less immediately.

In saying that it treats a person's productive capacity as something over which she enjoys a "natural privilege,"[22] Marx is characterizing the socialist proportionality principle as a truncated form of the principle of self-ownership. To be sure, he does not expressly acknowledge that the privilege of self-ownership is truncated here, but it clearly is, since the socialist principle demands what self-ownership would demand in a world of communally owned means of production only *after* deductions prejudicial to self-ownership have already been applied.[23]

Because it preserves the income entitlement of self-ownership, albeit in a restricted form, the socialist proportionality principle "is therefore a right of inequality": the more talented will, *pro tanto*, do better than the less talented, and those with few dependents will be able to sustain each of them better than those with many. But these departures from equality of provision—"these defects"—"are inevitable in the first phase of communist society as it is when it has just emerged after prolonged birth pangs from capitalist society. Right can never be higher than the economic structure of society and its cultural development conditioned thereby."[24] You cannot expect the principle of self-ownership, after centuries of sway, to lose all of its power on the very morrow of the socialist revolution. For practical reasons,[25] socialists must sustain the continued writ of self-owner-

22. It is hard to be sure what Marx means when he says that socialism "recognizes unequal individual endowment . . . as [a] . . . natural privilege." On an entirely literal reading of his statement, he means that the socialist regime affirms the piece of bourgeois ideology that says that people have fundamental moral rights of self-ownership. But it is more likely that Marx means that socialism treats personal endowment *as though* it were a natural privilege: not the ideology as such but the principle of distribution it mandates is preserved. There is, in any case, no need to decide between these two interpretations of Marx in the present essay.

23. One might question whether all of the deductions contradict self-ownership, but some of them certainly do.

24. Note, by the way, that the unequal provision is without any ado called a "defect." If, as some contend, Marx did not believe in equality of condition, why would he consider that a defect? (For a sophisticated presentation of the stated contention, see Allen Wood, "Marx and Equality," in *Analytical Marxism*, ed. John Roemer [Cambridge: Cambridge University Press, 1986]. For a refutation of Wood's contention, see Norman Geras, *Literature of Revolution* [London: Verso, 1986], pp. 48–51.)

25. The persistence of self-ownership (whether it be as ideology or just as practice: see note 22 above) under socialism is, in part, a hangover from the recent capitalist past: reward is still needed as an incentive to contribution because people are used to it. But Marx presumably also means here (in line with his historical materialism, as that theory is understood in my *Karl Marx's Theory of History* [Oxford: Oxford University Press, 1978]) that the bourgeois principle (or practice) is most suitable for advancing the development of the productive forces toward abundance in virtue of their immediately post-capitalist level, and not (only) because of a form of ideological lag. If he had thought that lag was the whole explanation of the defect in socialist justice, he would not have been so certain that the

ship, which is not to say that they must espouse it as a fundamental principle.

Well, how do we get beyond the defective—because still partly bourgeois—regime that prevails in this lower phase of communist society, socialism? You might think that we get beyond it by confronting the source of the inequality of condition that is the defect of that society, namely, that the society "tacitly recognizes unequal individual endowment and thus productive capacity as natural privileges." You might think that we get beyond it by repudiating the persisting legacy of the self-ownership principle. But no. Here is where Marx fails forthrightly to reject left-wing libertarianism. He sidesteps the issue by resorting to a technological fix. Having said that the defects of socialism are inevitable, given its emergence from capitalist society, he immediately goes on to write a paragraph that has inspired many people, including me, but about which I shall shortly make some complaints. Here is what he says:

> In a higher phase of communist society, after the enslaving subordination of the individual to the division of labour, and therewith also the antithesis between mental and physical labour, has vanished;[26] after labor has become not only a means of life but life's prime want; after the productive forces have also increased with the all-round development of the individual, and all the springs of co-operative wealth flow more abundantly—only then can the narrow horizon of bourgeois right be crossed in its entirety and society inscribe on its banners: From each according to his ability, to each according to his needs!

What communist society inscribes on its banners might seem to be inconsistent with libertarianism. Does it not imply that the individual must lend his ability to the community, for the sake of equality of condition, and therefore exercise it differently from how he might otherwise have chosen to do? But in communist society everyone develops freely and, so I infer, without any such constraint, and the reason why universal free development is compatible with equality of condition is that labor has become life's prime want: the uncon-

---

defect would persist until abundance had been achieved: a more perfect justice might have come in advance of abundance as bourgeois consciousness wore off.

26. With the vanishing of that antithesis, "labour in which a human being does what a thing could do has ceased": Karl Marx, *The Grundrisse* (Harmondsworth, Middlesex: Penguin Books, 1973), p. 325.

strained exercise of the ability of each not only allows but also promotes satisfaction of the needs of all. And the reason why labor has
become life's prime want is that it has become attractive, because of
the high level of productivity posited earlier in the paragraph: "the
material means of ennobling labor itself" is to hand.[27] Hence my
claim that Marx sidesteps rejection of left-wing libertarianism by resorting to a technological fix. At the high material level of abundance,
the incentive of reward for labor contribution is no longer necessary,
and so it is not now necessary to maintain the writ of self-ownership.
But the same thing that makes it unnecessary to maintain its writ
also makes it unnecessary to reject the principle itself. When labor is
life's prime want, it is performed both without the remuneration en-
joined by the principle of self-ownership and without the coercion
that presupposes rejection of that principle.

In both phases of communism people contribute whatever labor
they choose: that is why nothing is said about how much they should
contribute in the lower phase. Since labor is then their only means of
life, there is no danger that they will not contribute. In the higher
phase, labor, having become "ennobled," is now "life's prime want,"
so, again, there is no danger of deficient contribution. "From each
according to his ability" is not an imperative but part of communism's
self-description: given that labor is life's prime want, that is how
things go here. People fulfill themselves in work that they undertake
as a matter of unconditional preference rather than in obedience to
a binding rule.

IV

Marx's (higher phase of) communism is characterized by an egalitarian distribution that is not achieved by (the threat of) force: I shall
use the phrase "voluntary equality" to denote that feature. I shall
now ask, How might voluntary equality be possible, in a large modern society? and I shall treat Marx's own reply to that question (it is
possible when and because material abundance prevails) as only one
of the answers it might receive. Some people will think that a society
of voluntary equality is ipso facto a communist society, while others
will insist that, to be communist, a society must have further (for

27. Karl Marx, letter of March 9, 1854, in *The People's Paper*, in *Collected Works*, 13:58.

example, distinctive institutional) features. I do not take a stand here either on that question or on how Marx himself would have answered it, for my immediate topic is voluntary equality; I continue to discuss communism only because voluntary equality is (at least) a necessary condition of it.

I am not trying to show that society-wide voluntary equality *is* possible, though I do believe that it is. The question of how it might be possible is not the same as the question of how it is possible. The former question, as I mean it here, asks, What, most plausibly, would make voluntary equality possible, if it is possible? In what follows I discuss (and reject) what I have taken to be Marx's answer to that question: a different answer, which I endorse; and a third answer, often attributed to Marx, which I deplore. The two answers attributed (by me, and by others) to Marx are consistent with left-wing libertarianism. The answer I prefer is not.

In what I understand to be Marx's account of how voluntary equality is possible, a plenary abundance ensures extensive compatibility among the material interests of differently endowed people: that abundance eliminates the problem of justice, the need to decide who gets what at whose expense, and, a fortiori, the need to implement any such decisions by force. But while it might have been excusable, a hundred years ago and more, to ground the possibility of voluntary equality on an expectation of effectively limitless productive power, it is no longer realistic to think about the material situation of humanity in that pre-green fashion. The "springs of co-operative wealth" will probably never "flow" so "abundantly"[28] that no one will be under the necessity of abandoning or revising what she wants because of the wants of other people. The problem of justice will not go away.

Yet a lesser abundance might make the solution of voluntary equality possible, albeit in a different way. I have in mind a level of productive capacity that is too low to prevent significant conflicts of material interest from arising but high enough to enable their peaceful egalitarian resolution. For not all conflicts of interest generate a need for coercion. A conflict of interest obtains when there are things a person wants that she can get only if, as a result, someone else fails to get something she wants.[29] But a conflict of interest generates a need for

28. *Critique of the Gotha Programme*, p. 324.
29. Cf. Lars Bergström, "What is a Conflict of Interest?" *Journal of Peace Research* 3 (1970), 208.

coercion only when at least one party to the conflict is unwilling either to modify what she wants or to forbear from pursuing it unless forced to do so. I think that it is not utopian to speak of a level of material plenty that, while too low to abolish conflicts of interest as such, is high enough to allow their resolution without coercion and in favor of equality.[30] Notice that I do not have in mind here just any old peaceful compromise somewhere or other between opposed demands. I am speaking of an uncoerced *egalitarian* resolution of conflicts of interest, and I conceive such a resolution to be possible when and because people act from a belief in egalitarian justice that, material conditions being favorable, demands only some and not a heroic sacrifice of their self-regarding interests. I project such a society as feasible, not because I think that people might become *entirely* just, and therefore prepared to sacrifice their interests to whatever degree, in whatever circumstances, the virtue of justice would require, but because I think that they are or might become *sufficiently* just willingly to support an egalitarian distribution, in conditions of modest abundance. That is my solution to the problem of how voluntary equality might be possible. (I am concerned here with motivations, and I want to leave institutional details open, but let us suppose, for vividness, that equality is secured through a swingeingly redistributive egalitarian tax law, with nonpayment of the required tax generating no penalty except, perforce, disapproval.[31] But even that penalty is never applied, since everyone believes in paying, and willingly pays, the required tax.)

Someone might ask why I insist on considerable (even though not unlimited) abundance as a presupposition of voluntary equality. He might say that it is just as likely to be displayed under less bounteous and perhaps even harsh conditions, where it indeed demands more sacrifice from the talented, but where they might be moved by the greater (than in more benign conditions) liability to deprivation of the less able. Well, I do not need to reject a view about the material conditions of voluntary equality that is more optimistic than the one

30. I believe that per capita income in advanced capitalist societies is already beyond the required level, but that belief is not, of course, integral to the above claim.

31. Some might say that a law with the character of a primary rule (in Herbert Hart's sense) must have a penalty attached to it. But I do not think that is true, and even if it is, its truth is irrelevant here, for I could speak without loss of an accepted and recorded principle of redistribution and not call it a "law." (For a good discussion of whether laws require penalties, see Joseph Raz, *Practical Reason and Norms* [Oxford: Blackwell, 1975], pp. 157–62.)

I have affirmed. But anyone who is drawn to such a view should bear in mind that we are considering here the regular long-term operation of a large-scale, nonprimitive society that is not at war. This means that behavior characteristic of acute situations of shared distress, such as famine or disastrous accident, is irrelevant here, that familial and tribal forms of solidarity are here ruled out,[32] and that the civic virtues that go with war are not here to be expected. With all of those pertinent stipulations in place, it seems unrealistic to hope for voluntary equality in a society that is not rich. But, however that may be, I am chiefly concerned here to address not the optimists described above but Marxists who believe that substantial conflicts of interest make equality impossible, and I ask them to consider whether the material progress on which they rest their hope might not enable an egalitarian resolution of those conflicts at a stage earlier than the one at which that progress causes them to vanish.

V

It should be evident that Marx's account of the possibility of voluntary equality does not contradict left-wing libertarianism. And one might think that left-wing libertarians also need not object to the society my own solution describes, since it, too, displays no coercion invasive of self-ownership. But in my solution the absence of coercion is consistent with social equality only because the citizens disbelieve in self-ownership and willingly follow a law (or practice[33]) reflecting that disbelief. They believe that those who cannot produce as much as others are nevertheless entitled to comparable provision, and they consequently also believe that force should be imposed in (at least some) societies unlike their own, where force *is* necessary to promote greater equality. Because everyone has those beliefs and is true to them, the better endowed fulfill their equality-sustaining obligations without being forced to do so. And because the constitution of the society imposes no penalties to sustain equality, it is thus far acceptable to left-wing libertarianism. But the tenets of the society, which are what render coercive laws unnecessary, are not ones that a left-wing libertarian could endorse.

32. See further, *Karl Marx's Theory of History*, pp. 211–12.
33. One could rephrase everything that follows to satisfy the view of those with whom I expressed disagreement in note 31.

Left-wing libertarians will have stronger reservations still about such a society if its egalitarian law has penalties attached to it. One might think that, because such penalties would, *ex hypothesi*, be unnecessary as a means of ensuring that people act in the prescribed egalitarian way,[34] there could be no point in adding them. But a case for adding penalties that are never meted out might nevertheless be made.

The reason for having such penalties, their proponent urges, is to emphasize that infirm people (to take an extreme case of lesser ability) are not sustained because of the mere generosity of others. The penalties serve to underline the message that the infirm have rights: it is not just a failure of charity when the able do not succor them. Since infirm people have a right to support that is incompatible with granting to the able full rights of self-ownership, there should be a (hypothetically) coercing state, or state surrogate, that sustains their right to support, *even* when people would support them anyway. (A state surrogate is, for example, a general disposition among people to apply coercive sanctions against miscreants, without a special institutional apparatus for applying them.)

What the state here directs the able citizen to do conforms to her independently determined will, but she is still under threat of penalty if she does not do as it directs. It is, to be sure, a mistake to think that just because citizens would anyway follow the state's directives, there could be no reason for them to object to the threat attached to those directives. There are at least two reasons why people might justifiably feel aggrieved when forcibly obliged to do what they would do anyway. But neither applies in the present case.

The first reason is illustrated when a repressive state forbids people to visit countries to which, as it happens, they do not want to go. They might nevertheless quite reasonably resent the prohibition, since they might think that which countries they visit should be wholly up to them. But the citizens of the egalitarian society we are contemplating do not have this reason for resenting the fact that they would be forced to do what they anyhow want to do, since they recognize the justice of the infirm's claims and therefore agree that

34. I am here supposing that everyone knows that everyone knows (and so on) that everyone is a committed egalitarian. Since that is common knowledge, in David Lewis's sense (see *Convention* [Cambridge: Harvard University Press, 1969]), legal penalties are not even necessary for "assurance game" reasons. For the concept of an assurance game, see A. K. Sen, "Isolation, Assurance, and the Social Rate of Discount," *Quarterly Journal of Economics* 80 (1967), 112–24.

they should be forced to respect those claims of justice if, what is false, it is necessary to force them to do so.

A different reason for objecting to the existence of a penalty for doing what I have no intention of doing is that I might resent the lack of confidence in my moral sense which the existence of the penalty could be thought to betoken. But that ground of grievance is also absent here, for the penalties are not in place because of what able people might otherwise do. Their rationale is to reinforce the status conferred by the law on underendowed people. That rationale means that there might be a point in having the penalties even though everyone knows that in their absence no one would act differently. The penalties are not adopted because of adverse beliefs about anyone's motivation.

That concludes my defense of the claim that unused penalties would lend emphasis to the rights of less talented people. Whatever should be thought about that claim, I am confident that persons of left-wing disposition should reject Marx's utopian abundance as the basis for projecting a future social equality, and that they should ground the possibility of equality in a willingness, in the context of a lesser abundance, to act on beliefs about justice that are inconsistent with any form of libertarianism.

VI

I express optimism when I say that we may envisage a level of material plenty that falls short of the limitless conflicts-dissolving abundance projected by Marx, but which is abundant enough so that, although conflicts of interest persist, they can be resolved without the exercise of coercion. I think Marx himself would have been more pessimistic on that particular score. I think he believed that anything short of an abundance so fluent that it removes all major conflicts of interest would guarantee continued social strife, a "struggle for necessities . . . and all the old filthy business."[35] I think it was because Marx was needlessly pessimistic about the social consequences of anything less than limitless abundance that he needed to be so optimistic about the possibility of that abundance. A pessimism about social possibility helped to generate an optimism about material possibility.

35. *German Ideology*, p. 49.

Consider this statement from *The German Ideology:* "So long as the productive forces are still insufficiently developed to make competition superfluous, and therefore would give rise to competition over and over again, for so long the classes which are ruled would be wanting the impossible if they had the 'will' to abolish competition and with it the state and the law."[36] What can it mean for competition to be *superfluous,* as opposed to just nonexistent, nonactual? It can only mean, I suggest, that there is nothing for anyone to compete *about:* competition is superfluous when everyone can have everything she wants without prejudice to the wants of others.

Now, Marx says here that as long as competition is *not* superfluous, competition is "therefore" inevitable: competition, he says, is necessary as long as it is possible. But what makes competition inevitable, what "give[s] rise to it over and over again," is an insufficient development of the productive forces: only when the productive forces are sufficiently developed is competition superfluous and therefore only then is it possible to transcend it.

It follows from all this that, as long as it is possible not to have communism, communism is impossible, and that the only thing that can make it possible is a plenary development of the productive forces. Having burdened himself with that gloomy confidence, Marx had to believe either that capitalism would last forever or that such a development of the productive forces would one day come. Since he hated capitalism, he needed his technological fix.

VII

Some students of Marx would say that I have exaggerated the extent to which he relied for the success of communism on an abundance that would extinguish conflicts of interest, and I admit that I got the idea not just from poring over texts but from remembering how bourgeois objections to communism were handled in my childhood and youth. For I was brought up in the Canadian communist movement, and my first ingestion of Marxism, which no doubt had a lasting effect, was in that movement; it might be claimed that the emphasis on limitless abundance in my own understanding of Marxism comes not from Marx but from that movement. (The movement's

36. Ibid., pp. 329–30.

textbook was Stalin's *Dialectical and Historical Materialism:* for Joseph Vissarianovitch, the development of the productive forces was the key to history and to human liberation.)[37] The students of Marx I have in mind would agree that, according to Marx, there will be few conflicts of interest under communism, but they would say that he based that expectation *not* on belief in a future towering abundance but on confidence that, after the abolition of capitalism, individualist *motivation* will gradually wither away. People will no longer think in terms of mine and thine. Each person will become a "social individual" who *identifies* herself with the interests of other people.

I have reservations about this interpretation of Marx, which I express in section IX. Here I want to criticize reliance on the transcendence of individualist motivation, whether or not there is a basis for it in Marx. One can, of course, question whether such a transcendence is possible. Putting that aside, I want to urge that it is extremely undesirable. For unless the phrase "transcendence of individualist

---

37. To provide the reader with a taste of what Stalinist faith in the development of the productive forces was like, I quote from the Soviet textbook *Fundamentals of Marxism-Leninism,* which appeared in the early 1960s. Its account of what occurs in "the advance to the shining heights of communist civilization" begins with a long quotation from the work of academician V. A. Obruchev, "the well-known Soviet scientist," who, we proceed to learn, did not go far enough:

"It is necessary:
"to prolong man's life to 150–200 years on the average, to wipe out infectious diseases, to reduce non-infectious diseases to a minimum, to conquer old age and fatigue, to learn to restore life in case of untimely, accidental death;
"to place at the service of man all the forces of nature, the energy of the sun, the wind and subterranean heat, to apply atomic energy in industry, transport and construction, to learn how to store energy and transmit it, without wires, to any point;
"to predict and render completely harmless natural calamities: floods, hurricanes, volcanic eruptions, earthquakes;
"to produce in factories all the substances known on earth, up to most complex—protein—and also substances unknown in nature: harder than diamonds, more heat-resistant than fire-brick, more refractory than tungsten and osmium, more flexible than silk and more elastic than rubber;
"to evolve new breeds of animals and varieties of plants that grow more swiftly and yield more meat, milk, wool, grain, fruit, fibres, and wood for man's needs;
"to reduce, adapt for the needs of life and conquer unpromising areas, marshes, mountains, deserts, taiga, tundra, and perhaps even the sea bottom;
"to learn to control the weather, regulate the wind and heat, just as rivers are regulated now, to shift clouds at will, to arrange for rain or clear weather, snow or hot weather."
It goes without saying that even after coping with these magnificent and sweeping tasks, science will not have reached the limits of its potentialities. There is no limit, nor can there be any, to the inquiring human mind, to the striving of man to put the forces of nature at his service, to divine all nature's secrets. (Moscow: n.d., pp. 876–87)

motivation" is used as a misleading description of something different, to wit, my own idea that conflicts of interest persist but people are willing to resolve them in an egalitarian way, it denotes a pretty hair-raising prospect, in which self-ownership is protected at the expense of an oversocialization of the self-owning individuals. The Soviet jurist E. B. Pashukanis looked forward to the "social person of the future, who submerges his ego in the collective and finds the greatest satisfaction and the meaning of life in this act,"[38] but I do not find the ideal portrayed in that formulation attractive.

In the Pashukanis scenario, the constitution lets people behave as they please, "just as they have a mind," and what they are minded to do is to please each other. My own solution to the problem projects no such extreme change in people's motivation. *I think it is far better to reject the bourgeois normative principle of self-ownership and not look to a*

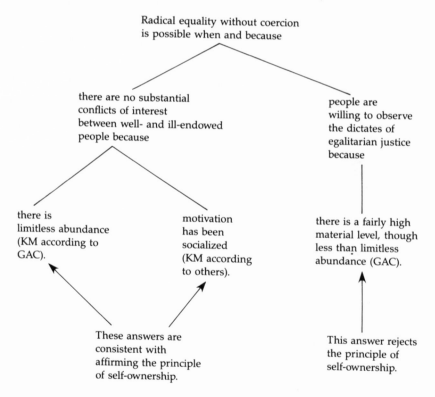

38. E. B. Pashukanis, *Law and Marxism* (London: Pluto Press, 1978), p. 160.

big transformation in human psychology than it is to defer to the bourgeois principle and rely on an extravagant degree of socialization.

The insistence on overcoming all conflicts of interests, in both its abundance and "social individual" versions, reflects an infantile unwillingness to countenance a measure of self-denial as a way of dealing with the inevitable difficulties of social existence. We should maturely accept that there will always be substantial conflicts of interest, but that people may be able to handle them with mutual forbearance and a sense of justice when they are blessed with material circumstances that are clement, yet not Elysian. For this to be feasible, people do not have to be zealously just and altruistic, since I am premising an abundance that, while smaller than what I think Marx prophesied, is great enough to ensure that very considerable self-sacrifice for the sake of equality of condition will not be necessary.

The diagram above summarizes much of the foregoing discussion. It depicts three bases for believing that radical equality without coercion is possible.[39]

VIII

The diagram displays two traditional Marxist answers to the question of how voluntary equality might be possible, and my own different answer. It omits a fourth answer, which is neither Marxist nor my own, which I shall call the Sumner solution, because Wayne Sumner suggested it to me.

The Sumner solution is unlike any Marxist one in that it tolerates conflicts of interest, but it is also unlike mine in that its mechanism is not a sense of egalitarian justice, and it therefore does not contradict libertarianism. The Sumner solution depends on a material level like the one I invoke, at which large sacrifices are not needed for equality to be achieved. Its achievement is based on a form of limited altruism. Under Sumner, people have sufficient (not unlimited) sympathy for one another, so that whenever anyone falls below the common level of well-being, others notice her plight and take remedial action. The assistance they give is not offered with a redolence of kindness, which could offend the dignity of underendowed people, but in a spirit of

39. For a challenging critique of an earlier version of the preceding sections of this paper, see Keith Graham's reply to my "Self-Ownership, Communism, and Equality," in *Proceedings of the Aristotelian Society, Supp. Vol.*, 1990.

comradeship. Sumner citizens are not Pashukanis people, since they do sacrifice goals they would wish to pursue, in the way that one might for the sake of a friend.

In the Sumner scenario, no one aims at equality as such, as she would if she were actuated by justice, but an egalitarian upshot is nevertheless achieved. I find that conjunction of conditions unrealistic, at any rate for societies larger than villages. It might be realistic to suppose that, even in a large society, people unmoved by justice could be firmly disposed to help anonymous others who are in special difficulty (other than in a spirit of charity), but the most that would bring about is an anarchist analogue of the welfare state: relief from serious trouble and acute deprivation. I do not see how the Sumner motivation could be expected to secure the more demanding goal of equality.

IX

I expressed doubt (p. 149) about the attribution to Marx of a Pashukanis-like view of what would make communist equality possible. Perhaps he relied on radical sociality at the time of his earlier writings, but socialized individuals, in the required sense, do not seem to me to be what makes communist equality possible in his later works, from *The German Ideology* on. When *The Communist Manifesto* predicts that "the free development of each [will be] the condition for the free development of all," it does not, so I have suggested (see section III above), forecast an access of altruism or any other similarly large change of affect. Let me revert to the music-making example. If a saxophonist, a bass player, a drummer, and one or two others form a jazz band because it is the heart's desire of each to develop her own talent by playing in one, they are not actuated by altruism, and no one joins the band in order to promote another's self-realization. Each joins it to fulfil herself, and not because she wants it to flourish for any independent reason (which is not to say that she does not also have such independent reasons).

In his *Marxism and Morality* Steven Lukes overlooks the jazz-band interpretation of the relevant texts and consequently embraces a strongly oversocialized and textually unsustainable characterization of the Marxist communist individual. I know no evidence in Marx's mature writings for Lukes's view that he "inclined towards the maximum" erasure of distinction between individual interests.[40] Communism is not, Marx said, the "love-imbued opposite of selfishness."[41]

In my view, when the mature Marx spoke of an overcoming of the antagonism between the individual and the social interest, he did not mean that the individual would make the social interest his own. What he had in mind was a happy harmony of individual interests and, consequently, a cessation of the expression of social unity in alienated forms.[42] Lukes misconstrues the absence of conflict between the individual and a *separate* social interest as an identity of the interests of individuals.[43] Marx did not, moreover, posit an absurdly complete harmony of individual interests. The antagonisms deriving from "the social process of production" have gone, and so, therefore, has the need to project an illusory harmony in the state and religion, but not all "individual antagonism" disappears.[44]

To be sure, Marx did use the phrase "social individual," and it suggests, out of context, the product of a radically socializing transformation in the structure of individual motivation. But a look at the passage in which the phrase occurs dispels that initial impression. When Marx affirms that, with communism, "the development of the social individual . . . appears as the great foundation-stone of production and of wealth,"[45] he is not talking about altruistic motivation or any other sort of affective bond. He is talking about sociality no longer being confined to an alienated expression in repressive state and social structures:[46] with the supersession of alienation, society now exists only in its constituent individuals.

In that way the "social individual" of the *Grundrisse* recalls the individual of *On the Jewish Question,* who will recognize and organize his own forces "as *social* forces, and consequently no longer separate

40. Steven Lukes, *Marxism and Morality* (Oxford: Oxford University Press, 1985), p. 98.
41. Marx and Engels, "Circular Letter against Kriege," in *Collected Works,* 6:41.
42. See my *Karl Marx's Theory of History: A Defence,* pp. 125–29, on how, according to Marx, antagonism between individuals leads to sociality in alienated forms.
43. See Lukes, *Marxism and Morality,* pp. 97–98. The passage from *German Ideology* that Lukes quotes at p. 98 has nothing to do with socializing the individual.
44. Preface to *The Critique of Political Economy,* in *Selected Works in One Volume,* p. 183.
45. *Grundrisse,* p. 705. For similar phrasing, see p. 832.
46. For this different interpretation of "social individuality," see further David Archard, "The Marxist Ethic of Self-Realisation," in *Moral Philosophy and Contemporary Problems,* ed. J. G. D. Evans (Cambridge: Cambridge University Press, 1987), pp. 28ff.; and Will Kymlicka, *Liberalism, Community, and Culture* (Oxford: Oxford University Press, 1989), pp. 114–19. The socialization of the individual, as interpreted by those authors, is related to the notion of communism as "the liberation of the content," which I explore in sec. 7, chap. 5 of *Karl Marx's Theory of History.*

social power from himself in the shape of *political* power."[47] But in that early work the idea that sociality as oppressive superstructure will subside *is* linked to a motivational doctrine. For when the self-emancipating individual of the *Jewish Question* transcends the state, she transcends, in the same movement, the egotism in civil society that made the state necessary. That is one reason why I apply only to Marx's mature writings my claim that he rejected the socialized individual (in the Pashukanis sense) as a basis for the possibility of communism. Another is the apparent textual support for the socialization notion in the extraordinarily ambitious description of what it is to "[carry] out production as human beings," to be found in Marx's "Comments on James Mill" of 1844.[48]

X

Does communism, as Marx depicts it, and given the nature of his reason for thinking it possible, realize justice, or is it, as some people think, "beyond justice"?[49] That question is different from two other ones, namely, Does communism, as Marx depicts it, realize justice, *on Marx's understanding of justice?* and, *Did Marx himself think* that communism would realize justice? I am asking only the first question, which is the least exegetical of the three. The three questions are plainly distinct and, in my view, have different answers, partly because, as I have argued elsewhere,[50] while Marx believed that capitalism was unjust and that communism was just, he did not always realize that he held those beliefs. This discrepancy within his attitude to justice has generated exegetical controversy, on which I do not comment here, partly because I should have so little to add to Norman Geras's excellent treatment of the issue.[51]

Part of the controversy with respect to whether or not Marx's communism is beyond justice reflects failure to distinguish categorially different subjects of the predicate "is just." One thing that can be

47. *Marx/Engels*, 3:168.
48. *Marx/Engels*, 3:227.
49. "Beyond justice" is multiply ambiguous. "Beyond" could mean "beyond the need for" or "beyond the possibility of," and "justice" can be a feature of persons, institutions, or distributions. I do not treat all of the resulting possible disambiguations in what follows.
50. In a review of Allen Wood's *Karl Marx*, in *Mind* 92 (July 1983), 440–45.
51. See "The Controversy about Marx and Justice," in Geras, *Literature of Revolution*, pp. 3–57.

just or unjust is the distribution of goods and services. I think that it is just when it has a certain egalitarian shape, whose exact character need not be specified here, but which I shall suppose characterizes distribution under Marxian communism. It would follow that communism, as Marx depicts it, indeed realizes justice. But it does so either (as in my interpretation of Marx) by virtue of abundance or (as in the interpretation of him offered by others) by virtue of a certain socialization of motivation. In neither case, then, is the justice it realizes due to justice as a moral virtue in its citizenry. And since one can think of justice not as a rightful state of affairs but as a way of achieving one, then, given how the rightful distribution is achieved in communism, as Marx depicts it, one may conclude that it is beyond justice. But there is no reason for someone who thinks that communism is just because its distribution is just to resist the conclusion that it is (also) beyond justice, in the sense in which I used "beyond justice" in the foregoing sentence. (Note that, under my own non-Marxist proposal about what might make communism possible, the society is in no sense beyond justice in either of the two senses distinguished above—justice as a distribution and justice as a moral virtue.)

The importance of being clear about what justice, in different discussions, is a property of is borne out by an interesting relationship between three views of what might make communism possible, which I have discussed in this essay, and certain distinctions in David Hume's account of the personal virtue of justice.

For Hume, one subjective condition and one objective condition are individually necessary and jointly sufficient for the virtue of justice to be both necessary and possible in human society. The conditions are limited generosity and moderate scarcity.[52] When limitations on generosity are removed, so that each person "feels no more concern for his own interest than for that of his fellows,"[53] justice is unneces-

---

52. ". . . 'tis only from the selfishness and confin'd generosity of men, along with the scanty provision nature has made for his wants, that justice derives its origin" (*A Treatise of Human Nature* [Oxford: Oxford University Press, 1888–], p. 495). For a well-argued claim that Hume's list is incomplete, see Lukes, *Marxism and Morality*, pp. 32–33, and also his "Taking Morality Seriously," in *Morality and Objectivity*, ed. Ted Honderich (London: Routledge and Kegan Paul, 1985), pp. 104–5. Lukes follows, and develops, Allen Buchanan, *Marx and Justice*, p. 167. For further sapient criticism of Hume, see James Griffin, *Well-Being* (Oxford: Oxford University Press, 1986), pp. 285–86, 382.

53. Hume, *An Enquiry concerning the Principles of Morals*, ed. P. H. Nidditch (Oxford: Oxford University Press, 1975), p. 185. Note that this is not the paradoxical altruism in which people care only about the interests of other people. It is, rather, an immediate attitude of conformity to the policy of either maximizing or equalizing welfare (depending on how the equality of concern is to be understood: see p. 157 below).

sary, and when people lack generosity entirely, justice is impossible. Under complete absence of scarcity, which is the same thing as unlimited abundance, justice is, once again, unnecessary, and under dire scarcity, where there is less than enough to sustain everyone, it is impossible: you cannot expect human beings to be just in such circumstances. Justice is a normal feature of human affairs, since its subjective condition is fulfilled as a matter of human nature, and its objective condition is almost always fulfilled, since dire scarcity cannot last (it wipes people out), and few people live on naturally bountiful South Sea islands.[54]

54. Here is a more systematic presentation of Hume's account of the conditions of justice. Generosity and scarcity each may be entirely absent, moderate, or unconfined. That creates nine possible conjunctions of objective and subjective conditions. I tabulate them here, together with some consequences I think Hume would draw, in each case, for whether the virtue of justice is possible and necessary:

GENEROSITY

S C		Absent	Moderate	Unconfined
A R	Unconfined (= dire scarcity)	1 Impossible	4 Impossible	7 Unnecessary
C I	Moderate (= moderate abundance)	2 Impossible	5 Possible and Necessary	8 Unnecessary
T Y	Absent (= unlimited abundance)	3 Unnecessary	6 Unnecessary	9 Unnecessary

Entries 3, 6, 7, 8, and 9 are justified by this *Treatise* sentence: "Encrease to a sufficient degree the benevolence of men, or the bounty of nature, and you render justice useless, by supplying its place with much nobler virtues, and more valuable blessings" (p. 495). And Hume's brilliant discussion in pt. 1 of sec. 3 of the *Enquiry* appears to me to justify the remaining entries in the table.

When scarcity is moderate, generosity makes justice possible, and its confined character makes justice necessary. When generosity exists but is confined, the absence of dire scarcity makes justice possible, and the presence of (moderate) scarcity makes it necessary.

Is justice possible in the cases where it is unnecessary? It seems to me impossible in cases 7 through 9, for the reasons given at pp. 158 below, where I discuss the inconsistency of increased benevolence and justice. But is justice possible in cases 3 and 6, where what makes it unnecessary is not special generosity but abundance alone? One might say: no, it is not possible, because abundance renders unnecessary creation of the institutions cleaving to which counts as justice, so they won't be there to be cloven to. But one might also say: yes, it is possible, since it is possible for people pointlessly to erect and honor those institutions; you could have property rights where, for reasons of abundance, they are unnecessary, so you could have pointless justice, "an idle ceremonial" (*Enquiry*, p. 184), in cases 3 and 6. Or should one still say: no, since

That is Hume's answer to the question, When is justice, as a virtue of persons, both necessary and possible? I asked a different question: What might make communism possible, where communism is conceived as egalitarian distribution without coercion? The questions are different, but the answers map on to one another. The first answer to my question is the one I think Marx gave, to wit, the circumstance of massive abundance, which is the same as one of Hume's two answers to the question, What would make justice unnecessary? The second answer to my question, which others think Marx gave, I called socialized motivation, and that is like unlimited benevolence, which is Hume's alternative answer to the question of what makes justice unnecessary. The answer I favor to my question is voluntarily just behavior. That is what Hume says occurs when people are neither wholly uncaring about others nor thoroughly mutually identified and nature neither imposes dire scarcity nor offers a cornucopia;[55] and in my account of communism, by contrast with the two accounts of it attributed to Marx, it, too, displays those subjective and objective features.

One might say, in Humean vein, that in the two Marxian views of it (as opposed to my own), communism is beyond justice because it is beyond one or other of its two conditions. Or one might, on the other hand, say that what, for Hume, makes justice unnecessary is precisely what, for Marx, makes justice possible. But the air of paradox disappears once we recall what justice, in the two cases, is a feature of.

If we predicate justice of communism, as Marx depicts it, then we do so because of its egalitarian distribution, not because of a virtue in its citizens. For Hume, by contrast, justice is always a virtue of persons. It is, among other things, and most relevantly here, the artificial virtue that restrains our natural tendency to appropriate what a generally welfare-promoting property institution assigns to others.

Both "generally" and "welfare-promoting" need explaining. The institution obedience to which is justice is only *generally* welfare-promoting because, in particular instances, adhering to the property

---

honoring those institutions would not go against the grain, so it would not be an artificial virtue, so it would not be justice?

55. Except that Hume and I regard very different things as the content of just behavior: the parallel is rather more formal at this point than it is at the other two.

rules can do more harm than good, and it is a mark of the artificiality of justice (by contrast with a natural virtue like benevolence) that this should be so. There are no feasible property rules that are better, all things considered, than one that in some instances enjoin something harmful.

The expression "welfare-promoting" is, by design, ambiguous across two alternatives. Some commentators think that Hume was a utilitarian, so that the institution obedience to which is justice promotes aggregate happiness, if necessary at the expense of the happiness of some. Others, like David Gauthier, think that Hume supposes that the relevant institutions promote the welfare of each person.[56] I use the vague term "welfare-promoting" because the difference between its aggregative and individualized interpretations is not important here.

Institutions like private property are not in themselves, for Hume, just or unjust but welfare-promoting or -frustrating; and people are just or unjust according to whether they do or do not observe the rules of welfare-promoting institutions. Justice is "a regard to the property of others," but that regard counts as justice only because the institution of property is generally welfare-promoting.[57]

Since Humean justice is adherence to the rules of property, there can be no justice, for Hume, where property rules do not exist. Accordingly, when full abundance makes those rules unnecessary, it also dissolves justice. Justice is the artificial virtue that combats our natural tendency to take what belongs to others. But there is no scope for the expression of that tendency when abundance obtains because human beings can then get all they want without taking from others (and they are not so malicious as to want to take from others for noninstrumental reasons). Abundance means that there is nothing for the artifice of justice to oppose.

It might seem less clear that increased benevolence will "render justice useless."[58] Why should a disposition to respect rules of property not coexist with an expansion of generosity? As long as there is scarcity, there seems to be a point in distinguishing mine from thine, and if, being benevolent, people observe the distinction willingly,

---

56. See Gauthier's excellent "David Hume, Contractarian," *Philosophical Review* 88 (January 1979), 3–38.

57. "Of the Original Contract," in *Hume's Moral and Political Philosophy*, ed. H. D. Aikin (New York: Hackett, 1959), p. 367. Sometimes keeping promises is *also* part of justice; see, e.g., Hume, *Treatise*, p. 526.

58. Hume, *Treatise*, p. 495.

then are they not being willingly just rather than benevolent-and-therefore-not-just?

That line of questioning mistakes what is involved when the limitations on human generosity are lifted. When each "feels no more concern for his own interest than for that of his fellows," the use of any material object on any occasion falls, with everyone's consent, to whomever would appear likely to get the most satisfaction from it.[59] There would, accordingly, be no "land-marks between my neighbour's field and mine," and not merely no walls or fences.[60] Its definitional association with property makes justice by nature a "cautious, jealous virtue," a regard for mine and thine ungoverned by immediate welfare consequences, and therefore strictly incompatible with the degree of generosity that Hume says banishes justice.[61]

If justice is a disposition to honor certain institutions and injustice is a disposition to violate them, then neither exists, we are truly beyond both, when the institutions in question do not exist. But if justice is regarded as a feature of distributions, then a variety of motives is consistent with justice, and also a variety of circumstances, including abundance. If justice is, for example, distribution proportional to need, then abundance might be thought to make justice easy to achieve rather than unnecessary. And where abundance is lacking, then special generosity, even Humean unlimited generosity, might be what makes justice possible. In the widest sense of distribution[62] there is always a distribution of goods and bads in society, and it always has a certain shape. Then where justice is a distribution of a certain shape, it is impossible to go "beyond" justice and injustice. They are contradictories, not merely contraries.

APPENDIX

Ronald Dworkin objects to my use of the concept of self-ownership to identify libertarianism.[63] He says that the idea of self-ownership is too indeterminate to pick out a distinct position in political philoso-

59. Or to whomever has some differently grounded welfare claim on it, depending on how equality of concern is understood: see note 56 above.
60. Hume, *Enquiry*, p. 185.
61. Ibid., p. 184.
62. Wider, for example, than one that restricts itself to distributions of property per se.
63. The objection I describe was raised by Dworkin at a seminar in Oxford in the summer of 1986.

phy. He reasons as follows: to own something is to enjoy some or other set of rights with respect to that thing. But one might envisage a number of importantly different sets of rights over themselves and their own powers in virtue of which we could say of people that they are self-owners. The principle of self-ownership, as it was described above, therefore lacks determinate content.

The premises of this skeptical argument do not appear to me to sustain its conclusion. They do not show that the principle of self-ownership legislates indeterminately. For they do not refute the hypothesis, which I hereby propose, that the principle achieves determinacy through requiring that *everyone* enjoy *full* self-ownership rights. It might indeed make no determinate sense to say, in the abstract, that Jones owns herself. But when one says that each person has full private property in herself, then the constraints of universality and fullness combine to limit and perhaps to determine precisely what rights each person must, on pain of inconsistency, have over herself. My right to use my fist as I please must, under universal complete self-ownership, stop at the tip of your nose because of your rights, under universal complete self-ownership, over your nose.

The principle of self-ownership confers the fullest right a person (logically) can have over himself provided that every other person also has just such a right. That statement generates a procedure for determining the content of the self-ownership principle. Determinacy might be achieved through the identification of a set of rights S that every person can have over herself and where S confers fuller rights over oneself than does any other set of universally enjoyable rights. Dworkin's premises do not show that there is no such set, and I believe that there is one. But even if I am wrong about that, and no set meets the stated condition, my guess is that the plurality of maximal rights-sets would not differ in ways that matter to the central questions in political philosophy on which the thesis of self-ownership bears. Since, for example, failing to help another person could not constitute an interference with her right to use herself as she wishes, and not being required to help others leaves one with more rights over one's powers than one would otherwise have, the right not to supply a service or product to anyone else will form part of any plausible reading of the self-ownership principle.

Self-ownership is the principle that, to adopt a phrase of Dworkin's, allows "differences in ability" to "produce income differences in a laissez-faire economy among people with the same ambitions."[64] Laissez-faire economies can vary in their rules of operation, but that is not a reason for condemning the unexplicated idea of a laissez-faire economy (which Dworkin uses) as hopelessly indeterminate. The idea of a person's ownership of herself might be capable of different interpretations, but it is no less determinate than the unexplained idea of her ownership of other resources, to which Dworkin qualmlessly and innocently helps himself in his own writing.[65] I do not think that it is harder to say what rights I have if I own myself than it is to say what rights I have if I own a tract of land, or a horse.[66]

64. Ronald Dworkin, "Equality of Resources," *Philosophy and Public Affairs* 10 (Fall 1981), 311.

65. See ibid., p. 283.

66. I thank David Archard, Annette Barnes, Diemut Bubeck, Alan Carling, Andrew Chitty, Jon Elster, Michael Gillespie, Keith Graham, Les Jacobs, Will Kymlicka, Andrew Levine, Chris Morris, John Rawls, Seana Shiffrin, Wayne Sumner, Jeremy Waldron, and Andrew Williams for their helpful comments on drafts of this essay. And I am deeply grateful to Arnold Zuboff, who supplied dozens of excellent criticisms and suggestions for improvement.

Suggestions for Further Reading

The locus classicus for the technological construal of Marx's historical materialism is his 1859 Preface to *A Contribution to the Critique of Political Economy*. See also Part 1 of *The German Ideology* (1846) by Marx and Engels.

Vigorous support for this construal of Marx is offered in G. A. Cohen, *Karl Marx's Theory of History: A Defence* (Princeton: Princeton University Press, 1978); John McMurtry, *The Structure of Marx's World-View* (Princeton: Princeton University Press, 1978); William H. Shaw, *Marx's Theory of History* (Stanford: Stanford University Press, 1978); and Allen Wood, *Karl Marx* (London: Routledge and Kegan Paul, 1981). For excellent criticism of this point of view, see Joshua Cohen, review of G. A. Cohen, *Karl Marx's Theory of History*, in *Journal of Philosophy* 79, no. 5 (1982), and Richard Miller, *Analysing Marx* (Princeton: Princeton University Press, 1984). There are replies to J. Cohen and Miller in G. A. Cohen, *History, Labour, and Freedom* (Oxford: Oxford University Press, 1988), chap. 5, and chaps. 1–9 of that book contain further relevant discussion. For a sophisticated treatment of the contending positions, see Jon Elster, *Making Sense of Marx* (Cambridge: Cambridge University Press, 1985), chap. 5.

For earlier presentations of Marx within a technological perspective, see G. P. Plekhanov, *The Development of the Monist View of History* (1895; Moscow: Foreign Languages Publishing House, 1956); Nikolai Bukharin, *Historical Materialism* (1921; Ann Arbor: University of Michigan Press, 1969); Vernon Venable, *Human Nature: The Marxian View* (New York: Knopf, 1945); Jakob Hommes, *Der Technische Eros* (Freiburg: Herder, 1955); and Alfred Schmidt, *The Concept of Nature in Marx* (1962; London: New Left Books, 1971).

7 Reason in Exile: Critical Theory and Technological Society

SHELDON S. WOLIN

I

The present age is continually being reminded that, like late nineteenth-century Vienna or pre–World War I Europe, it is experiencing its own fin de siècle. Thoughtful writers tell us that we are living through a time of endings—of American hegemony, of history, of Marxism, of philosophy, and of modernity.

Although in each of these examples the meaning of "end" is significantly different, there is a common element of anxiety hinting that some distinctively modern enterprises are in extremis. One tempting interpretation, prompted by the collapse of communism abroad and the triumph of American conservatism at home, is that we are witnessing the end of deep crisis; that the last major turning point has occurred; that the old formula of "crisis = opportunity" has been fulfilled so that opportunity, released from its dependence upon crisis, now points toward infinity. With ends no longer a matter of contestation, human beings can dwell forever in a kingdom of pure means.

Marxian political theory was the last representation of the ancient theme that deep crisis is theory's opportunity to change the world.

162

Many of its earlier critics had claimed that this belief was the spawn-
ing ground for the "totalitarian temptation" to shape society as
though it were potter's clay. Ironically, the theme appears to have
reemerged among the apologists for capitalism. Opportunity's new
formula, touted as the universal panacea for Asia, Central Europe,
the Soviet Union, as well as the rest of the world, is "high technology
+ the free market." Thus at the very moment when theory in the
form of a Marxist "utopia" is pronounced dead, theory in an equally
doctrinaire, but more economistic, form is being thrust upon those
societies struggling for relief from theory. The utopia of opportunity,
like the workers' utopia it has replaced, eerily repeats the same demand: present generations must sacrifice themselves for a future society whose assurance lies in the power of technology.

Assuming that a more skeptical interpretation is still credible, one
might appropriately consider two writers, Max Horkheimer and
Theodor Adorno, who nearly a half-century ago set down a very
different version of Endgame in which the opportunity created by
crisis produced a bad infinity whose formula was "total domination
through technology." The crisis involved the destruction in Germany,
Italy, Spain, and France of the type of liberal societies being promoted
today in Central Europe and Russia and the establishment of societies
of varying degrees of totalitarianism. Although Horkheimer and
Adorno were deeply critical of all dictatorships, including Stalin's,
they believed that the crisis could not be understood as the break-
down of parliamentary structures or the triumph of mass parties
organized around a charismatic leader. The crisis centered, instead,
in the culture of the intellect and involved the core notions in the
heritage of the Enlightenment: faith in progress and its major driving
forces, reason and science. The tangible expression of that faith and
of its power was modern technology in its various forms of mass-
production techniques, a vast array of consumer goods, rapid transportation of goods, services, and people, and, last but not least, mass
communications, including unprecedented technologies for dissemi-
nating education and culture. The faith in technology was especially
strong among Marxists, whose hopes for human emancipation had
been pinned to the productive potential of new technologies.

Confronting a world in which cruel dictatorships had come to
power in some of the world's most "advanced" countries, Horkheimer
and Adorno argued that the brute facts of the twentieth century—
totalitarianism, racism, death camps, genocide, cultural barbarism,

and the celebration of military aggression—made it no longer possible to believe in the inevitability of progress or the neutral character of modern technology. "The basis on which technology acquires power over society is the power of those whose economic hold over society is greatest. A technological rationale is the rationale of domination itself. It is the coercive nature of society alienated from itself."[1]

By itself that conclusion was not particularly original. What was striking was the claim accompanying it: that reason and science, the fundamental principles of "the party of humanity," were complicit in the horror of a "relapse into darkest barbarism."[2] Reason and science were not simply instruments exploited by totalitarian regimes but symptoms of a mind-set that was continuous with such regimes and set mind against itself. "Enlightenment is totalitarian": that, *in nuce,* was the thesis propounded by Horkheimer and Adorno in *The Dialectic of Enlightenment.*[3]

To appreciate the scope of their claim, it is necessary to realize that they intended the term "Enlightenment" to have a far broader reference than the eighteenth-century intellectual movement associated with the French philosophes and Kant. They traced the course of Enlightenment back to antiquity and forward to modern science and to the civilization shaped by capitalism and the cultural values of the bourgeoisie.

Today's reader, one with a postmodern sensibility, for instance, would probably be skeptical of an account whose plausibility seemed to depend upon an exaggerated notion of the power of reason to influence events. Admittedly, German scientists and philosophers

1. Max Horkheimer and Theodor Adorno, *The Dialectic of Enlightenment,* trans. John Cumming (1947; London: Allen Lane, 1973), p. 121. In this essay I do not attempt to treat Horkheimer and Adorno as representatives of the Frankfurt School. I make occasional reference to Herbert Marcuse, who, although a longtime participant in the work of the Institute for Social Research, the School's formal organization, later developed a distinctive political and theoretical orientation that, in crucial respects, departed from Horkheimer and Adorno and took a more radical turn than Jürgen Habermas. Obviously space does not permit me to refine the numerous distinctions or to enter the necessary qualifications regarding a group of thinkers who were as individualistic as they were sophisticated. The various differences are recounted in detail in Martin Jay's standard work, *The Dialectical Imagination: A History of the Frankfurt School and the Institute of Social Research, 1923–1950* (Boston: Little, Brown, 1973).

2. Horkheimer, "Traditional and Critical Theory," in *Critical Theory: Selected Essays,* trans. Matthew J. O'Connell et al. (New York: Herder and Herder, 1972), p. 241.

3. *Dialectic,* p. 6. The work was written in 1944 but not published until 1947. It is critically but sympathetically considered by Jürgen Habermas in "The Entwinement of Myth and Enlightenment: Max Horkheimer and Theodor Adorno," in *The Philosophical Discourse of Modernity: Twelve Lectures,* trans. Frederick Lawrence (Cambridge: MIT Press, 1987), pp. 106–30.

probably contributed to the atrocities and aggressions associated with Nazi Germany, that reader might say, but those were individual transgressions and therefore individual actors rather than abstract nouns should be held accountable. Wary of reifications, a postmodern intellectual would doubt that any "cause of reason" existed to be betrayed even by time-serving Nazi intellectuals. More than likely she would be impatient with claims to objectivity of the sort which pre-postmodernity had associated with reason and science. The postmodern is inclined, instead, to a deflated view of reason, treating it as a convention of discourse at best and at worst a rhetorical strategy.

That response, while attractive for its modesty, may be using unpretentiousness as a strategy for deflecting some sensitive questions about the role of the philosophical intellectual. At the height of the Cold War all sorts of people worried that the future of the world might be decided by the use or nonuse of atomic weapons. Perhaps more public intellectuals should have been troubled by the fact that both major world powers employed professional reasoners who were able to give coherent and consistent accounts of why one side should or shouldn't drop the bomb, in retaliation or preemptively, upon the other. Perhaps in a world where peace depended upon the logic of mutually assured destruction (MAD), reason had already been transformed by technical/technocratic requirements.

The transformation might be reconstructed as the double institutionalization of reason. In a technocratic society even reason requires credentials. It must first be certified as professionally qualified by an institution of higher learning, preferably a "research university," as, for example, a doctor of rational choice or a game theorist. Next, reason must be adapted to the culture of government bureaucracies and of quasi-governmental, quasi-corporate think tanks and quasi-academic "centers." Reason emerges outfitted with a distinctive praxis: philosophy as the policy science of the new kingdom of means. The form that reason thus acquires, the political and social locations where it is employed, and the masters that it serves are precisely those which Horkheimer and Adorno had anticipated. "Since Descartes bourgeois philosophy has been a single attempt to make knowledge serve the dominant means of production, broken through only by Hegel and his kind."[4]

The postmodern response—if, indeed, that is what I have been

4. Cited in Jay, *Dialectical Imagination*, p. 8.

describing—tends to evade the problems initially posed by Hork-heimer and Adorno and restated later by Herbert Marcuse. Perhaps this is because the postmodern suspicion of metanarratives disables it from fully comprehending the shock and disillusionment produced among intellectuals by the collapse of the old humanistic cultures of Germany and Italy into totalitarian regimes and by the transforma-tion of Marxism into Stalinism. Or perhaps the events following the dramatic overthrow of communist regimes have so persuaded intel-lectuals that liberal capitalism is the true embodiment of the Eternal Return that they can say, along with Richard Rorty, that "Western social and political thought may have had the last *conceptual* revolution it needs."[5] Or perhaps it is because postmoderns have tacitly accepted a role that combines powerlessness and complicity and thereby shields the intellectual from any deep feelings of individual or corpo-rate responsibility toward the systems of power shaping their society. Where Hobbes and Locke, for example, had conceived reason as the attribute of an individual, indeed, all individuals, the postmodern looks for reason in the "guidelines" prescribed by prestigious institu-tions (for example, corporate and governmental foundations and aca-demic institutes) and experiences individual reason as institutional recognition/reward for "creativity" or "originality."

II

How was it that in enlightened times unenlightened despotism was possible? that modern despotism was able to incorporate the sci-ences, the arts, and the professoriat? that intellectuals who had once struggled against the premodern tyrannies now mostly served the interests of avowedly anti-intellectual regimes? that people of reason had come to betray civilized values and the promise of modern sci-ence and industry to abolish misery?

The fate of reason in the modern world, Horkheimer and Adorno claimed, could be understood as an act of "self-destruction," long in the making, by the bearers of the Enlightenment rather than simply the result of terrorism and regimentation of the sciences and arts by fascism and communism.[6]

5. Richard Rorty, *Contingency, Irony, Solidarity* (Cambridge: Cambridge University Press, 1989), p. 63.
6. *Dialectic*, p. xiii.

Horkheimer and Adorno proposed a genealogy of reason whose effect was, like Nietzsche's *Genealogy of Morals,* to unsettle philosophy by exposing its complicity with the forces of antiphilosophy. Unlike Nietzsche, however, they sought to retain some semblance of the "traditional" meaning of reason and of philosophy. In pursuing that purpose, they would take a most un-Nietzschean turn toward a will to powerlessness. Determined to rescue theory from power and to preserve a critical distance, they would pose the question of whether the disavowal of power might be a necessary precondition for critical theorizing.

The end point of this particular theoretical journey would be registered in Adorno's later formulation, "The moment to realize [philosophy] was missed . . . the attempt to change the world miscarried."[7] The political isolation and weakness of the representatives of theoretical reason was a major theme for Adorno and Horkheimer and in pursuing it they would come to reflect it, even consolidate it. The story they would construct was intended to make intelligible the experience of their times—the era of totalitarianism. It told of a deep, and perhaps fatal, diremption within reason itself. In their account, reason comes to confront its other, but that other turns out to be not unreason but reason itself. Although Horkheimer and Adorno would describe that confrontation as primarily a matter of opposing conceptions of philosophy/theory, their analysis was equally concerned with some of the oldest themes in the history of Western political philosophy/theory: how theoretical knowledge is transformed into power by submitting to certain self-imposed operations that enable it to contribute to the generation of power and the exercise of social control; and why theoretical reason is drawn toward total power, that is, toward an ideal society in which all human activities—from sexual to commercial to intellectual intercourse—are prearranged by reason and enforced by power so that they all "contribute" to some postulated end of goodness, well-being, justice, or piety.

Before I discuss the Frankfurt genealogy of reason and the account of reason's diremption, it is useful to recall that, historically, philosophy's self-appointed role as the guardian of reason was accompanied by sweeping political claims that established the nexus for the technical discussions of the grounds of reason in epistemology, logic,

---

7. Adorno, *Negative Dialectics,* trans. E. B. Ashton (1966; New York: Seabury Press, 1973), p. 3.

and cosmology/theology. What follows is a brief sketch of the early politics of philosophical reason.

III

Throughout most of its history Western philosophy has adopted reason as its principal totem. It was the so-called faculty of reason that was said to distinguish men from beasts, and even from women and children, as well as from other men who happened to be slaves or idiots. Early on, reason was identified as the peculiar province of philosophy and the special property of philosophers. Other men might have wealth, noble lineage, beauty, or great skill at some worldly occupation, such as fighting, governing, painting, or making shoes, but philosophers announced that what distinguished them from other human beings was that they either possessed or practiced reason or were single-mindedly devoted to acquiring it.

At the same time that philosophers were claiming a special relationship to reason and disagreeing among themselves about its right description, they mostly identified reason with truth, or, more precisely, with universal truths, that is, with truths independent of local time and place and powerful enough to bring harmony to the society that ordered its life in accordance with them. Philosophers assigned to themselves the responsibility of interpreting universal truths.

Within that broad economy of norms, political philosophers developed the more specialized claim that philosophical reason entitled them to judge society. The basis of that claim was the philosopher's alleged "knowledge of the whole." A political society was said to consist of "parts" (for example, classes, occupations, and functions such as war-making). The knowledge of how those parts should be interrelated and arranged so as to serve the "good of all" was a special knowledge that rulers needed but only philosophers possessed. "The law is not concerned with the special happiness of any class in the state but is trying to produce this condition in the city as a whole, harmonizing and adapting the citizens to one another by persuasion and compulsion."[8]

Thus it could be said that, historically, there is a standing quarrel

8. Plato, *The Republic*, trans. Paul Shorey (Cambridge: Harvard University Press, 1935), 7:519e.

between reason and politics. Political philosophers and theorists constructed political and social paradigms, allegedly on the basis of reason, to serve as models. The extreme case was the construction of "utopias," a word that, significantly, is translated as "nowhere" but could more literally be translated "no place" (topos = place), that is, a construction of reason, a complete whole, universally true because it occupies a timeless space disconnected from any historically configured place. But whether engaged in constructing utopias or developing more sober, political sciencelike constitutional typologies, political philosophers were mimicking in thought what rulers were bent upon doing in practice—ordering and regulating society, shaping it to the needs of power and, perhaps, justice, treating it as a whole made up of parts in need of coordination.

Because the theorist enjoyed greater freedom of construction than real rulers, he could use that advantage to judge existing societies and, almost without exception, he indicted them as unjust, unvirtuous, impious, or benighted. Each of these charges was inspired by the belief that insofar as a society was deficient in justice or piety, it was a defective whole and hence must be deficient in reason. Philosophers could not conceive of societies that were unjust or unvirtuous but rational or, conversely, that were just and virtuous though irrational.[9] And they firmly believed that for societies to be just and rational they had to be ordered totally toward those ends. From Plato to Marx, political philosophers identified the just and virtuous totality with the "realization" of philosophy or, in Plato's formulation, with the philosopher achieving full stature. The salvation of society could not be achieved without the fulfillment of philosophy and the ending of its political alienation. That state of affairs can be described as the rationalization of power and the empowerment of theory. Those who ruled would look to reason, and those who reasoned would look to rule, directly or indirectly.

IV

Horkheimer and Adorno engaged a range of questions very different from the ones posed by most ancient and modern theorists. Theorists had asked, What kind of regime would most effectively serve

9. There have been exceptions, of course, e.g., Mandeville and Montesquieu.

philosophy so that the latter might inscribe its truths upon society? Horkheimer and Adorno were concerned to inquire into the terms that reason/philosophy would have to inscribe upon itself in order to realize its perennial dreams of power. What kind of discipline, what sort of self-denials would it have to observe in order to serve as a decisive element in the structure of power?

Their answer was that reason would have to offer itself in the service of calculation, self-interest, and self-preservation.[10] It would also have to narrow the range of reason, discarding the centuries-old idea that the truths of reason comprehended and determined the validity of moral beliefs and that philosophy was the final arbiter concerning competing claims as to the meaning of the good life for human beings. Reason would disavow the preoccupation with ends or a summum bonum and turn against that enterprise.[11] It would declare that ends were endless, never final, but tentative, experimental, or perhaps simply emotive and subjective. It would focus upon means and techniques of rational inquiry and rational decision. It would become a technology that Horkheimer and Adorno christened "instrumental reason" and, as such, reason and its theoreticians would cease to be powerless. The latter would be enrolled in a structure of power that was peculiarly modern, both in its exploitation of science for the development of technology for the needs of production and for its totalizing power representing the domination of nature and society by those who controlled economic power.[12]

Horkheimer and Adorno set the problem of reason in a context that was richly suggestive of what we might call archetypal themes in the history of political theory, concerning the relationship of reason to society, of philosophy to reason and to the dominant political powers, and of philosophical reason to political context. In their most influential work, *The Dialectic of Enlightenment*, the emergence of rea-

10. *Dialectic*, pp. 86, 88, 91.
11. Ibid., pp. 92-93.
12. Ibid., p. 121. The theorist who more than anyone else developed the concept of instrumental rationality was Max Weber, who was also one of its first critics. Although Weber expressed ambivalence about the progressive elimination of different modes of relating to the world—religious, mystical, metaphysical, and heroic—he could no more renounce instrumentalism than he could capitalism, though he might denounce them. Both were extraordinary forms of power, and modernity, whatever else it might mean, was committed to the project of developing power, not abdicating it. See his two classic essays, "Science as Vocation" and "Politics as a Vocation," in *From Max Weber: Essays in Sociology*, ed. Hans Gerth and C. Wright Mills (New York: Oxford University Press, 1946), pp. 77–156; and my essay "Max Weber: Legitimation, Method, and the Politics of Theory," *Political Theory* 9 (1981), 401–24.

son in the West is described by a complex interpretation of Homer's *Odyssey.*[13]

Odysseus's encounters with various mythic powers are interpreted as reason struggling to disengage itself from myth but, as we shall see, only half succeeding. The encounters occur in the course of Odysseus's wanderings before the gods will permit him to return to his homeland after the Trojan Wars. Homeland, which also has mythical connotations, stands for reason's return from the world of myth.

The relationship between reason and homeland takes on a special poignancy for Horkheimer and Adorno. They, along with most of the other original members of the Frankfurt school, were involuntary exiles from their native Germany, having fled to avoid the fate of millions of other Jews during the Nazi tyranny. The Nazi suppression of philosophy as a free activity was accompanied by official encouragement of various doctrines that pre-postmodernity had no difficulty in designating as irrational myths: of *Blut und Boden, Lebensraum,* a pure Aryan *Volk,* and an inspired *Führer.* But, as the Horkheimer-Adorno construction of the myth of Odysseus attempted to show, reason was infected by myth, yet it could coexist with, even make use of it. Nazism may have suppressed genuine philosophy, but it incorporated reason in its modes of organization, its techniques of mass manipulation, and its development of military strategies. Thus philosophy went into exile, but reason remained at home.

The history of political theory is rich in examples of theorists who have been exiled from their homelands. One thinks of Hobbes fleeing the civil war and revolutionary upheavals of his day; of Locke forced to go abroad because of his close ties with the Shaftesbury opposition to the Stuarts; or of Marx journeying to London in the aftermath of 1848 to escape the police of several countries. Unreason, in their eyes, had taken possession of their homelands. Their exile might be interpreted as the forced substitution of reason for homeland—Plato's "city laid up in heaven." All returned in a literal but not necessarily a theoretical sense, for the dream of philosophy had been to make reason and homeland synonymous. Locke may be said to have fully returned, but not Hobbes and Marx. We may want to pose the ques-

13. Parenthetically, although Horkheimer and Adorno did not make the point, it was common for Odysseus to be treated as a symbol of philosophy. See, for example, Thomas More, *Utopia,* bk. 1, 11, p. 49; *Complete Works,* vol. 4, ed. Edward Surtz and J. H. Hexter (New Haven: Yale University Press, 1965).

tion, does continued exile, whether actual or not, cause philosophy to adopt mythical elements as it seeks to preserve its dream of being transformed into power? We shall come back to that question later.

It is now possible to see that the diremption of reason into instrumental and critical reason becomes the means of salvaging a rational element for philosophy, that is, of preserving some version of a union of reason and philosophy. But the difficulty is that philosophy has lost its homeland because instrumental reason has been installed there. Philosophy, in the form of critical reason, is then condemned to permanent exile, to wandering as Odysseus had, to engaging mythical powers and becoming infected by myth but doomed to powerlessness. To wander is to have no context. We might think of exile, therefore, as signifying a certain kind of loss, loss of political context. In a later section I show that the crucial and defining element of theoretical context is political and that philosophical reason has had a troubled relationship with political context.

Turning now to a more detailed examination of instrumental reason, I want to explore both the self-inscribed conditions that philosophy/reason adopts in order to become power and, accordingly, the kind of power that philosophy qua instrumental reason thereby becomes.

V

Latter-day instrumental reason appears, in its self-understanding, as a combination of pragmatism and scientific empiricism, of practical applications founded upon verifiable knowledge. The undeniable triumphs of modern science and the dazzling technologies they have made possible encourage a belief that instrumental reason is the discovery of the true nature of reason and of its only legitimate form.

Critical theory attempts to go behind that claim and expose a mixed genealogy. It proposes to reveal how, through various encounters with the powers of society and nature, instrumental reason has acquired elements of antireason. Reason's genealogy is linked to Greek myths and the Book of Genesis to establish that, as soon as it begins to extricate itself from myth, reason reproduces mythic elements. It gains knowledge of how to preserve the self, how to calculate its interests, but it is unaware that it is mutilating the self, making of it a sacrificial object to the cult of power.[14] Accordingly: "If by enlighten-

---

14. *Dialectic*, pp. 54, 86, 88, 91.

ment and intellectual progress we mean the freeing of man from superstitious belief in evil forces, demons and fairies, in blind fate— in short, the emancipation from fear—then denunciation of what is currently called reason is the greatest service reason can render."[15]

Throughout its career reason undergoes various transformations yet never completely expels myth, and in the end the two have become inseparable. Reason, in the practice of positivist science and logical positivist philosophy, reverts to an uncritical attitude that renders its own assumptions mythic even as it declaims the more strongly its fidelity to fact and "reality." For the essence of mythology is the acceptance of the status quo: "cycle, fate, and domination of the world reflected as the truth and deprived of hope." Modern positivist science, with its relentless emphasis upon shaping thought so as to enable it to grasp the facts better, its reproduction of the immediacy of existing facts in the guise of mathematical abstractions, confirms the existing state of things. "The world as a gigantic analytic judgment . . . is of the same mold as the cosmic myth which associated the cycle of spring and autumn with the kidnapping of Persephone." The scientific emphasis upon recurrence and regularities reproduces fatefulness and reveals how, in the guise of discovering the "necessary" laws of nature, "enlightenment returns to mythology."[16]

Although historically instrumental reason first appears in antiquity, its full flowering occurs in the modern era when science and industry are constituted by the dynamic needs of capitalism. It operates from within the assumptions of a social order organized by modern capitalism's conception of production and is structured, therefore, according to a logic in which all relationships are commodified and conceived as exchange relationships. It is, however, a special mark of those who practice instrumental reason that they are largely unaware of how it has been socially constituted, especially in its scientific form. Accordingly, critical theory attacked what it conceived to be the myth of the neutral, impartial observer/investigator, the nonreflexive subject who is trained to divest the self of all cultural influences (= "biases") save those of scientific education. In reality, science and empiricism have produced a socially and politically unconscious being, unaware of whom he is serving or of whom she is helping to exploit.

Instrumental reason is a peculiar combination of passivity and aggressiveness. Politically it is deferential, content in its subordination to political and economic powers, but as applied knowledge or prac-

15. Horkheimer, cited in Jay, *Dialectical Imagination*, p. 253.
16. *Dialectic*, pp. 26–27.

tice it displays a different face. It attacks the natural world as a domain to be conquered and a storehouse of utilities to be exploited. The human world is treated as continuous with the natural and hence as a population of manipulable objects fit for social engineering. Reason's instrumental character is expressed in its aspiration to make the most efficient and effective use of what is in the world within the framework and according to goals defined by the ruling authorities.

Instrumental reason is more than a conception of reason as means. It is an antipolitical way of developing power in the handling of objects, animate as well as inanimate, that qualifies as despotic. Objectification is not so much an epistemological category as the expression of a power impulse. If objects are to be used efficiently and effectively, they must be simplified by being represented solely in terms of features or traits that render them manipulable or exploitable. Accordingly, empirical objects, including human beings, whose varied and variable natural or acquired elements of identity resist use and exploitation, must first be stripped of their culturally constituted identities and then reinstated as abstract entities that are now suited to display the regularity, recurrence, and uniformity requisite for classification. Their histories and biographies are irrelevant to their social and political uses.

As reason becomes increasingly methodized and systematized, philosophy becomes mimetic of what it believes to be science. It becomes dismissive of metaphysical claims and views ethical concerns as being expressive of emotions rather than reason. Truth acquires a literalness, and all dialectical notions concerning the contradictory character of reality are ruled out.[17] As reason becomes radically narrower, a more or less technical mode of discourse identified with various specializations, its scope of application grows ever larger. In the twentieth century it is embodied in and supports a system of universal domination, "an administered world," ruled by "state capitalism."[18] Thus the paradox: as reason grows narrower, more dogmatic, its practical applications multiply exponentially as capitalist societies expand their power over their members toward an ideal limit of totalization in which all human relationships are continuously

17. For a later restatement of the position see Adorno's Introduction in Theodor W. Adorno et al., *The Positivist Dispute in German Sociology*, ed. Glyn Adey and David Frisby (1969; New York: Harper & Row, 1976), pp. 34–36.

18. *Dialectic*, p. x. The seminal essay on that concept was Friedrich Pollock's "State Capitalism: Its Possibilities and Limitations," in *The Essential Frankfurt School Reader*, ed. Andrew Arato and Eike Gebhardt (New York: Urizen Books, 1978), pp. 71–94.

scrutinized, corrected, and coordinated. "The totalitarian order marks the leap from the indirect to direct forms of domination while still maintaining a system of private enterprise."[19] As reason moves inexorably toward its totalization as technical knowledge, in effect, it leaves no room for critical reason. The philosopher is thus doubly exiled, from both real and ideal homeland.

VI

In a famous essay, "Traditional and Critical Theory" (1937), Max Horkheimer expanded the idea of instrumental rationality into a full-fledged theoretical conception, maliciously dubbed "traditional theory" to emphasize its association with the status quo. Modeled primarily after the dominant logical positivist conception of science, it can be summarized as the reconstitution of reason into a technical tool with a consequent shedding of a critical and substantive/legislative role.

In Horkheimer's account, traditional theory proceeds as though there is nothing problematic in its fundamental notions. For example, it recognizes a "fact" as something that is a "datum" or "there" or "empirically proven." This might be described as "the myth of the given." Instrumental reason dismisses as trivial the complaint that its abstractions are created by wrenching some event or phenomenon from its context, that in ignoring, say, the historical forces, social interests, and ideologies that had shaped a particular fact, reason was not only distorting the fact in question but reconstituting it. Reason proceeds this way because of its role within a system of production. Facts need to be smoothed of their peculiarities so that they may be reintroduced as uniform, regular, or homogenous. Thus processed, facts can be more easily assembled to generate lawlike statements that, in turn, present a manipulable world where practical applications and predictions become a matter of course. Thus epistemic modes mimic capitalist modes: just as capitalism subordinates the worker's personal needs to the objective requirements of production, so the scientist applies reductionist methods to facts in order to dissolve their individual peculiarities and force them to conform to, that is, work on behalf of, some hypothesis.

19. Horkheimer, "The End of Reason," in *Frankfurt School Reader,* p. 34.

At one time reason had operated as a critical, even a liberating force, opposing ignorance, superstition, and oppression. But it also undermined the grounds on which, historically, human beings had established value and meaning. Religion was condemned as superstition, ethical and political norms were said to be mostly metaphysical and hence without rational warrant, while the idea of nature as a distinctive realm of value was dismissed as sentimentalism and replaced by a notion of nature as an experimental domain of lawlike regularities that served as the basis of scientific predictions and control.

Having demonstrated that competing modes of knowledge are unscientific alternatives and hence powerless, reason becomes the sole oracle to be consulted when choosing among various means. The criterion that governs choice is, What will be the most effective/efficient way to a particular goal? A rational society would be one in which this mode of thinking was extended to all significant areas. In Horkheimer's formulation, "its features can be summarized as the optimum adaptation of means to ends, thinking as an energy-conserving operation. It is a pragmatic instrument oriented to expediency, cold and sober."[20]

By virtue of its association with science, logical positivism shares vicariously in power. In its explication of instrumental reason, in its verification principle, which dismisses as "metaphysical" or "emotive" all statements that are not scientifically verifiable, logical positivism reveals itself to be the theoretical counterpart to fascism. In insisting that philosophy and science rid its language of all traces of subjectivity, metaphysics, and value, logical positivism displays "the need for a purge [*Sauberung*] to which authoritarian states make the most appalling concessions."[21] Logical positivism allegedly "parallel[s] an attitude to culture which now and then finds practical expression in nationalistic uprisings and the bonfires associated with them."[22]

VII

"Critical theory" was the alternative that Horkheimer developed in the hope of retaining links between philosophy and reason while

20. Ibid., p. 28.
21. Horkheimer, "The Latest Attack on Metaphysics" (1937), in *Critical Theory*, p. 179.
22. Ibid., p. 174.

distinguishing critical from instrumental reason. "It is a critique of philosophy, and therefore refuses to abandon philosophy."[23] The crucial link that joined reason and philosophy and separated critical from instrumental reason was a claim that critical reason/philosophy was the true heir to a tradition in which the ideal of theorizing was inseparable from the idea of a social whole. The attack upon positivism, which Horkheimer and Adorno carried on for more than three decades, depended upon the validity of that idea. They defended critical reason as a superior understanding of the world that instrumental reason was helping to create but could not comprehend. The theoretical superiority of critical theory was attributed to a conception of reason as dialectical in character, grounded in historical self-consciousness, and able to grasp the social totality in all of its interconnectedness, including the connection between positive science and the dominant powers in society.

The theoretical power of critical reason is realized in "comprehensive theoretical statements." "In order to understand the significance of facts or of science generally one must possess the key to the historical situation, the right social theory."[24] In its reliance upon general principles and "conceptual wholes," the critical character of theory was inseparable from a conception of theory as privileged knowledge of social totality.[25] Even in Adorno's postwar writings, with their frequent disclaimers that concepts could be adequate to reality, there are assertions that "theory seeks to give a name to what secretly holds the machinery together."[26]

Critical theory's dialectical mode of thinking reinforced its holism. The contradictions exposed by that were significant because they were systemic.[27] "Within dialectical theory such individual facts always appear in a definite connection which enters into every concept and which seeks to reflect reality in its totality."[28] For the prewar Horkheimer and Adorno, dialectical history meant primarily a Marxian view of the evolution of society in terms of the division of labor,

23. *Dialectic*, p. x.
24. "The Latest Attack on Metaphysics," p. 159.
25. Horkheimer, "Traditional and Critical Theory," pp. 191, 218, and *Eclipse of Reason* (1947; New York: Seabury Press, 1974), p. 8. Substantially the same position was taken later by Adorno, "Sociology and Empirical Research," p. 69. See also Martin Jay, *Marxism and Totality: The Positivist Dispute in German Sociology* (Berkeley: University of California Press, 1984), chap. 8.
26. "Sociology and Empirical Research," p. 68.
27. See Herbert Marcuse, "A Note on Dialectic," in *Frankfurt School Reader*, pp. 444–51.
28. "The Latest Attack on Metaphysics," p. 161.

commodified exchange relations, and the domination of society by representatives of those who controlled the means of production.[29] Other Marxian categories, such as class struggle, the role of the proletariat, and the necessity of revolution, became, as I explain, steadily more problematic in the Frankfurt reading.

VIII

However superior critical theory might be because of its grasp of totality, there was a troubling quality about the notion itself. If totality was a fact of capitalist and totalitarian societies, and if critical theory claimed to be able to expose the true nature of that totality, could critical theory be critical without setting out its idea of a "true" totality? And what kind of politics was implied by a commitment to theoretical totality? The issue was not whether critical theory had a utopian vision of its own but whether totality was a proper theoretical project or simply the perpetuation of the mimetic of ruling that theorists had reproduced since antiquity, which suddenly had become ominous because totality was political fact rather than fantasy.

Curiously, Adorno had expressed deep reservations about the concept of totality before he and Horkheimer were forced into exile. His inaugural lecture at the University of Frankfurt in 1931 begins: "Whoever chooses philosophy as a profession today must first reject the illusion that earlier philosophical enterprises began with: that the power of thought is sufficient to grasp the totality of the real."[30] Reason, he argued, cannot discover itself in reality because reality has become hostile to reason. Philosophy itself faces "liquidation" as the sciences, logic, and mathematics carve out their respective domains and dismiss as insignificant what cannot pass scientific scrutiny. If philosophy is to be preserved in a fragmented world, it must concentrate upon "interpretation" rather than "research."[31] Philosophy "must give up the great problems, the size of which once hoped to guarantee the totality" and turn instead to historically concrete questions located in a "text" that is "incomplete, contradictory and fragmentary." Adorno likens philosophy to "riddle-solving" that does not seek to get "behind" the riddle but to "negate" it so that the

29. Horkheimer, "On the Problem of Truth" (1935), in *Frankfurt School Reader*, pp. 433–34.
30. Adorno, "The Actuality of Philosophy," *Telos* 31 (1977), 120.
31. Ibid., p. 126.

original question "disappears." Thus the riddle of the essence of commodities is exploded once commodities are analyzed in terms of their place within the structure of exchange.

Theory, Adorno seemed to suggest, could dispense with totality yet continue to guide praxis. Interpretations of "given reality" lead to the "abolition" of that reality. A praxis, he insisted, "follows promptly" from "the annihilation of the question."[32]

Was that formulation, however, an example of what Adorno would later call mythical thinking? One question had not been annihilated: Whose praxis? Horkheimer would write, "The methods, categories as well as the transformation of the theory" require the "taking of sides." "Right thinking depends as much on right willing as right willing on right thinking."[33] Taking sides presented no difficulty to orthodox Marxism. The working class was the chosen instrument of revolutionary praxis, of action whose ultimate objective was the overthrow of the capitalist system. In this formula, action and theory were symmetrical. For each the object was a totality. Theory explained the interconnected nature of the whole system from base to superstructure; revolutionary action would seize power and abolish the old values and institutions and completely reorganize social production.

Totalitarianism rendered that whole schema invalid. Not only had it made the idea of totality repugnant by representing itself as the realization of that idea, but it had incorporated and neutralized the working class. Theory seemed without either subject or object.

IX

Despite Adorno's philosophical misgivings about totality, he and Horkheimer clung throughout the 1930s to a Marxist perspective and its version of social totality. That decision enabled critical reason to avoid the possibility that its own distinction between instrumental and critical reason was a contrast between theory as power and theory as powerlessness. Marx had claimed that his theory was not only aligned with the direction of the most powerful forces ever amassed by human beings—forces that were rapidly converting the entire world into one economic totality—but that theory had discovered in

32. Ibid., p. 129.
33. "The Latest Attack on Metaphysics," p. 162.

the proletariat the power that was destined to reclaim that productive potential in the name of humanity. The inevitability of a proletarian revolution was linked by Marx to his critique of the laws of political economy.

Once Horkheimer and Adorno came to doubt that either capitalist exploitation or totalitarian oppression would provoke the workers to rebel, a wide range of issues opened up. Perhaps political economy was no longer an adequate description of the site of dialectical tensions and the workers no longer the instrument of revolutionary hopes. Where, if anywhere, was the possibility of revolution located, and who, if anyone, was to be its agent? Or were neither agency nor revolution any longer meaningful? Was theory now required to bear the entire burden of opposition? And if so, in what sense, if any, could theory profess to be a representation of power or to be connected in any meaningful way with social forces predisposed to the cause of radical change? "The theoretician . . . can find himself in opposition to views prevailing even among the proletariat."[34]

While Adorno, in order to survive, ironically found himself compelled to learn the language and practice of positivist social science, Horkheimer attempted to formulate a conception of theory in which the theorist seems peculiarly qualified by his isolation to undertake the role of opposition.[35] Opposition, he wrote, begins naturally: "an active individual of sound common sense perceives the sordid state of the world" and decides to change it. That desire not to submit to things as they are causes him to organize facts, to "shape them into a theory," thereby asserting that the subjective interests of the theorist are "an inherent factor of knowledge," not a disqualification.[36]

Those heroic words, while redolent of power images, are mythic in tendency, as is shown by the remarkable fantasy that follows. Horkheimer describes an imaginary society in which logical positivism and scientific empiricism are associated with a ruthless regime. Because of their view of facts, scientists are inhibited from giving any account of the significance of what is happening to them; they can only describe. If the society were then freed by the efforts of "active groups and individuals" who stood "in a different relation to theory," whose

34. "Traditional and Critical Theory," p. 221.

35. Adorno's involvement in the Princeton Radio Project headed by Paul Lazarsfeld is described by Martin Jay, *Adorno* (Cambridge: Harvard University Press, 1984), p. 34, and by Gillian Rose, *The Melancholy Science: An Introduction to the Thought of Theodor W. Adorno* (New York: Columbia University Press, 1978), pp. 97–98.

36. Horkheimer, "The Latest Attack on Metaphysics," pp. 161–63.

"specific action was contained in their very mode of perception," who, "fastening their eyes on a better life . . . saw through it all," only then would science "admit" that its former consciousness had been false.

Horkheimer then proceeds to fashion a tableau of ultimate vindication in which theory squares its accounts with the proletariat. He pictures the masses confessing their errors: "Indeed, the masses themselves would now realize that what they had formerly said and done, and even what they had thought in secret were perverted and untrue."[37] After castigating the masses for their "blindness" and "passivity" under Nazism, he concludes: "When the thoughtless crowd is mad, thoughtless philosophy cannot be sane."[38]

When Horkheimer, in a major essay of 1937, outlined the essential features of critical theory, he deliberately distanced it from the proletariat. There must be tension and conflict, he argued, between the theorist and "the most advanced sectors of the class . . . his thinking is to serve." The critical theorist, rather than the proletariat, assumes the role of a saving remnant: "In the general historical upheaval the truth may reside with numerically small groups of men. History teaches us that such groups, hardly noticed by those opposed to the status quo, outlawed but imperturbable, may at the decisive moment become the leaders because of their deeper insight. . . . The future of humanity depends on the existence today of the critical attitude."[39]

Adorno would supply the epitaph to theory's alliance with the workers. It read simply: "opposition to production."[40] The break with the proletariat also meant the loss of a political context, the context of revolutionary politics as defined by Marx and his successors. Adorno and Horkheimer refused to recontextualize their politics around the fate of democratic politics. They made no effort to develop a theoretical defense of even the troubled form of it in the liberal regime of Weimar or, later, the more robust social democratic politics of New Deal America.[41]

Declaring that "the might of industrial society is lodged in men's

37. Ibid., p. 161.
38. Ibid., pp. 167–68.
39. "Traditional and Critical Theory," p. 215. See Horkheimer's 1968 preface to *Critical Theory*, pp. v–viii, for his rejection of both the proletariat and the student left. In the writings of Marcuse and Habermas the proletariat would have all but disappeared without a trace.
40. "Traditional and Critical Theory," pp. 241, 242.
41. Adorno, *Minima Moralia: Reflections from Damaged Life*, trans. E. F. M. Jephcott (London: New Left Books, 1951), p. 15.

minds," Horkheimer and Adorno, especially the latter, began to explore the various ways in which popular culture was being manipulated to serve the ends of domination.[42] Adorno's picture of the masses reflects a mood of betrayal: they are portrayed as passive and, literally, mindless. There is not the trace of a suggestion that some popular cultures exist independent of mass audiences and may even be oppositional. There is, instead, a ressentiment against the masses, as though philosophy were chagrined at having been duped by Marxism into romanticizing the virtue of the oppressed. It was harshly expressed in a pensée by Adorno titled "They, the People": "But the moment simple folk are forced to brawl among themselves for their portion of the social product, their envy and spite surpass anything seen among literati or musical directors."[43]

Cultural critique, whatever its insights into the shaping of mass consciousness, proved a weak replacement for Marx's political economy. Marx's analysis, whatever its shortcomings, gives a plausible motive for the working class to rise up against its oppressors: misery, impoverishment, and the mechanization of labor are credible reasons for revolt. Cultural critique, in contrast, depicts an empty vessel, masses with no culture to lose, no motive for revolt. Critical reason has culminated in a theory about a nonrevolutionary subject.

X

"The truth becomes clearly evident in the person of the theoretician."[44] Horkheimer's brave formula represented an epitaph to philosophy's attempt to retrieve a measure of reason and power, but, in reality, critical theory ends in self-confessed powerlessness, surrendering one purchase point after another. It began by identifying critical theory with a power to grasp wholes; it ended with Adorno renouncing collectivity altogether—"the whole is the untrue"—and seeking to defend what was particular, individual, and, above all, negative against "totalitarian unity" in the postwar world.[45] Once it had declared for the close union of theory and practice, but later it

---

42. For an example of Adorno's lukewarm attitude toward the "political facade" of "late capitalism," see *Minima Moralia*, pp. 112–13.

43. See Jay, *Adorno*, chap. 4; Rose, *Melancholy Science*, chap. 6.

44. "Traditional and Critical Theory," p. 216.

45. *Minima Moralia*, p. 28.

gave up on the idea of developing a social praxis for fear that it would lead to embracing instrumentalism. The more isolated theory became, the more defiant its assertions of superiority over praxis. In Marcuse's words, "Theory will preserve the truth even if revolutionary practice deviates from its proper path. Practice follows the truth, not vice versa."[46]

In the postwar world critical theory would "demand support for the residues of freedom, and for tendencies toward true humanism," but that ultimatum was issued without regard to any political context and in the knowledge that these "residues" and "tendencies seem powerless in regard to the main course of history."[47] Philosophy is declared saved precisely because it has failed to be realized. As Adorno's famous epigram states, "philosophy, which once seemed obsolete, lives on because the moment to realize it was missed."[48] Theory becomes "thought as action," its own homeland, its praxis devoted to eluding dialectical completion.[49]

XI

Horkheimer and Adorno returned in triumph. They were welcomed by the postwar West German government, and their institute was reestablished to general and warm acclaim. Yet there is also a sense in which they, Adorno especially, never fully returned. In *Minima Moralia*, subtitled "Reflections from Damaged Life," Adorno referred to the "intellectual in migration" as if describing a permanent condition.[50] Although Nazi totalitarianism had been defeated, they felt that domination by state capitalism was as strong as ever, if not stronger. The Second World War seemed less a victory of liberal democracy than the repulse of the first primitive reconnaissance of instrumentalist totality. Where Nazism had exploited technology in the interests of propaganda, capitalism was using it to fashion a popular culture that would reinforce the psychology of domination without the rough hand of terror. "The development toward total integration

46. *Minima Moralia*, p. 18; Jay, *Dialectical Imagination*, p. 277.
47. Cited by Jay, *Dialectical Imagination*, p. 79. See also *Negative Dialectics*, p. 144.
48. *Dialectic*, pp. ix–x.
49. *Negative Dialectics*, p. 3.
50. *Minima Moralia*, pp. 148–51.

recognized in this book has been interrupted but not abrogated. It threatens to advance beyond dictatorships and wars."[51]

As the influence of technology was clamped more heavily on the culture of mass society and as other intellectual disciplines rushed to imitate natural science and exploit its prestige, the presence of scientific reason seemed ubiquitous in public and private bureaucracies, academic institutions, think tanks, and private life. While reason had thus become more firmly incorporated into the status quo and could rule by police but without a police state, the life of the mind remained in exile. The shape of reason, as Adorno saw it, had to be a "negative dialectics," its task to defend "difference," individuality, and what remained of high culture.[52]

That philosophy should find uncongenial and threatening a world dominated by reason, especially scientific reason, seems ironical and is a reminder of the tense relationship that has existed historically between philosophical reason and politics. In retrospect, *The Dialectic of Enlightenment* and other works by Adorno and Horkheimer appear as an extended meditation on philosophy's experience of exile and inevitably evoke the memory of Socrates, philosophy's martyr to politics. While the condemnation of Socrates by democratic Athens might symbolize the rejection of philosophy by the political, the decision of Socrates to refuse exile from the city and instead to submit to its judgment seems a puzzling choice. Did it signify the submission of the political to philosophy, or was it a statement about the tensions?

XII

Recall that in the *Crito* Socrates has been sentenced to death for subverting the laws of Athens, encouraging atheism, and corrupting the youth of the city.[53] He is entreated by his friends to escape punishment by going abroad, where he could continue to practice philosophy. In his response Socrates does not defend philosophy against the city but rather defends the city, arguing against his friends who regard the judgment of the city as inferior to the value of perpetuating philosophy. Socrates says he will not leave the city, which has nurtured him, nor subvert its laws by evading the judgment of the court. He not only refuses the life of exile but, in a pointed reference to the

51. *Dialectic*, p. x (from the 1969 preface). See also Rose, *Melancholy Science*, pp. 147–48.
52. *Minima Moralia*, pp. 148–51.
53. See John Wallach's fine essay "Socratic Citizenship," *History of Political Thought* 9 (1988), 393–413.

association of theorizing and traveling or journeying, proudly recalls that he has never left the city, save for military service.[54]

When Socrates defended himself before the Athenian democracy, according to Plato's *Apology*, he defiantly declared that, if given a choice of dying or desisting from philosophizing in public or of going into exile in another city, he would choose death. One reason Socrates advances for his refusal is that the god has "attached me to this city to perform the office of gadfly."[55] What makes Socrates' claim to a public office so striking is that it is accompanied by repeated disclaimers of ever having taken part in the politics of the city. The claim points to a kind of political practice proper to theory that is not the practice of politics but nonetheless explicitly presupposes it.

Socrates' refusal raises the question, if philosophy is identified with the rational pursuit of truth, why cannot the philosopher be anywhere as long as he is not persecuted? If reason is the quest for what is universally true, isn't that pursuit indifferent to context and hence essentially apolitical? Or is political theorizing different from philosophizing precisely because it needs a local politics, a homeland?

Let me interpret Socrates as saying that exile is the worst of fates for a theorist because it decontextualizes theoretical activity, freeing it of political ties and referents. Exile is analogous to the abstract form that theory assumes when it disavows a historically specific context. Because the exile is forced to become an alien in the double sense of being alienated from a "native" land and feeling alienated in a foreign land, she would have to resume theory in a "nonplace," exiled from the civic understandings that supply frames of reference to theoretical expressions. Theorizing requires a civic setting, not just any place but a particular one to which theorizing is attached by deep ties that, in the fundamental sense, are political without being mere politics. As Socrates' formula, it is to care more for "the polis itself" than for "the things of the polis."[56] That is, the ties are those defining and perpetuating a culture of shared understandings and practices that is no one's property and everyone's contribution. It is created by a variety of callings and skills and sustained by an implicit goodwill among the members. At the same time, however, a civic community, if it is to survive, must develop a structure of power and with that comes the likelihood of abuse and misuse of power.

In such a civic context the "office" of theory is, as Socrates claimed,

---

54. *Crito*, 51cff.
55. Plato, *Apology,* 31a.
56. Ibid., 36c.

to serve as the gadfly or critic of the dominant structures of power. The objective is not to criticize actors or policies, at least not directly, but to examine the lives of individuals because those lives embody cultural norms that indirectly constitute the politics of the society and contextualize it. The theorist cannot adopt the politics of those establishments without betraying his vocation. And this must not be done because power shares with reason the same temptation of wanting to be rid of the constraints of context: power wants to be omnipotent, reason wants to be omniscient.

On this reading, theory is dependent on a preexisting political context, and it is unnatural for it not to be. The unnatural is the abstract. It is unnatural because it entices thought into a realm where it is beholden to nothing except its own rules. Context is the conceptual trace of civic membership; it is theorizing as citizenship. It means not just objectively taking account of things or practices that have diverse origins and competing rationales but subjectively responding to them. Context works to soften, to tone down, even to exact an element of humility from the self, which recognizes that it is a multiplicity of debts to others, including those it could never have known. The recognition of context is political reason honoring its debts.

Instrumental reason, in the form of technology, is impatient with context and strives to be independent. The extraordinary achievements of technology have elevated one type of reasoning to archetypal status. Technology stands not just as new inventions or processes but for a culture of technical competencies. It is a culture that claims to be both *technē* and *logos*, skill and reason. Contemporary politics, with its reliance upon the mass media, media consultants, and pollsters; contemporary attempts to reduce politics to policy and to make policy the subject of the technical reasoning of economists, philosophers of public policy, and theorists of business management have all contributed to the depreciation and dismissal of more modest, more democratic notions of *technē* and *logos*. The fatal turn was present from the beginning.

One of the most striking passages in the *Apology*, hurried past by many readers, is where Socrates recalls how he had questioned a number of Athenians reputed to be wise or knowledgeable—politicians, sophists, poets, and persons of "great reputation." He concluded that "others who were supposed to be their inferiors were much better qualified in practical intelligence." Socrates then says

that finally he turned to "skilled craftsmen" and was not disap-
pointed. They proved to have a genuine kind of wisdom.[57]

In praising the practical intelligence of social inferiors and the wisdom of skilled craftsmen, Socrates was, I am suggesting, paying tribute to the importance of forms of competence that embody reason,
not pure reason but reason as culturally embedded: making a shoe
in Appalachia is not like making a shoe in Beverly Hills. But he was
also saying something as striking in its way as Plato's very un-Socratic
formula about philosophers as kings. He was calling attention to the
competence of ordinary beings, the praxis that arises from reflection
on experience, that is transmitted by a culture, and that represents
the power resident in the technologies of everyday life.

However, no sooner had Socrates made this discovery than he limited its importance by saying that, although the skilled craftsmen
possessed a form of genuine knowledge, they wrongly insisted upon
applying it to other areas where they lacked competence. Clearly
Socrates was alluding to the politics of democratic Athens, where, in
the public assemblies and law courts, ordinary citizens pronounced
on the weightiest of public issues. In Plato's *Republic* we know that
such intrusions into the political councils of the city were to be forbidden. That general prohibition against political participation by the
demos was not the simple conclusion to a claim about the superiority
of expert knowledge over alleged ignorance. It was a political strategy
by which philosophy chose to ally itself with kings and aristocrats,
the most powerful elements in the community. That choice meant
siding with exclusionary forms of politics and using philosophy's
claim to know the true foundation of political life to justify an even
more exclusive politics, so exclusive that politics disappears from
Plato's city. The philosopher had chosen to destroy politics to save
philosophy/reason.

As is usually the case, the best critic of Plato is Plato. In the *Protagoras* the sophist Protagoras defends the proposition that the skills
needed to run a city are part of a human capability for sharing in
justice and a moral sense. These can be acquired, Protagoras claimed,
through "deliberate choice, and by practice and teaching." It, too, is
a *technē* that can be taught and learned. And when it is, Protagoras
insisted, then "your fellow citizens are right to accept the advice of

57. Ibid., 216–22e.

smiths and cobblers on political matters." Without it, he declares, there is no sharing and hence no political association.[58]

Technē, we might say, is thus not the monopoly of technology or other technical applications of reason. The human capability for acquiring skills is not like the philosopher's claim to knowledge, the basis for a claim to power that excludes other claims. It is, instead, a claim that people cannot share in justice unless they share power. In our own day Protagoras's case has, if anything, gained even greater urgency. The growth of state power in advanced technological societies has been accompanied and promoted by the claims of technical reason, as in rational choice theory and its related methodologies and in the bureaucratization of expertise. These developments represent an attempt to make political *technē* the monopoly of governing elites while reducing the possibilities of a more democratized *technē* to the form of hiring its own technical reasoners to oppose governmental and corporate experts.

The danger to the political life of society posed by the dominance of a technological order does not come primarily from the incorporation of reason into the structure of that order, but rather from the incorporation of the philosopher. By "philosopher" I mean anyone whose vocation is the critical nurturing of public values. One of the virtues of democracy is that philosophers haven't liked it. Until late modernity, no defense of democracy was written by any philosopher or political theorist. The reason is apparent: the tradition of philosophy has wanted power to privilege reason. That tradition is now riding high, as philosophers, their protégés, and progeny come to occupy prominent places in public life. It is democratic reason, not philosophical reason, that is in opposition.

58. *Protagoras*, 324c–d.

Suggestions for Further Reading

Benhabib, Seyla. *Critique, Norm, and Utopia*. New York: Columbia University Press, 1986.
Dubiel, Helmut. *Theory and Politics: Studies in the Development of Critical Theory*. Cambridge: MIT Press, 1985.

Held, David. *Introduction to Critical Theory: Horkheimer to Habermas.* Berkeley: University of California Press, 1980.

Jay, Martin. *Marxism and Totality: The Adventures of a Concept from Lukács to Habermas.* Berkeley: University of California Press, 1984.

8 Technicity, Topology, Tragedy: Heidegger on "That Which Saves" in the Global Reach

REINER SCHÜRMANN

> The technicity of modern technique follows from a mu-
> tation in the essential way to be of truth. [. . .] From
> *alétheia's* conflictual way springs the possibility and the
> necessity of "tragedy."
>
> —M. HEIDEGGER, *Parmenides*

The organizers of the symposium at the same time raised and chased the specter of what one may call technological essentialism. In so doing, they placed themselves in good company. Michel Foucault wrote a *History of Madness* and later a *History of Sexuality*, only to tell us that neither 'madness' nor 'sexuality' designated enduring essences but rather problems that have their dates of emergence and their age of pertinence.

Likewise our hosts. They raised that specter inasmuch as the general title, "The Problem of Technology in the Western Tradition," suggests that we were invited to wrestle with a unitary issue that may have assumed various forms over the ages. In that, they have conventional wisdom on their side. Common sense would indeed have no qualms about distinguishing *kinds* of technology, for instance, pertaining to, or at least contemporaneous with—I quote from the titles of essays in this series—classical thought, Christianity, modern philosophy, Romanticism, and American liberalism. We do speak of Stone Age technology and DNA technology. One is then perfectly justified in suggesting that technologies come in the plural, that is, in various species subsumable under one generic concept. At the same time, however, we were there to discuss the technological project for the control and mastery of nature, clearly an exclusively modern project. Here the problem of technology is pertinent to one age alone. Control and mastery as projected by Galileo and Newton are not pertinent to the premodern world. The caveman's club has its *specific* technology; so do the medieval crossbow and today's MX missile. But the modern quest for mastery does not *specify* technology. When Descartes conceived a method "to render ourselves masters of nature,"[1] and inasmuch as modern technology realizes that project, one is clearly no longer speaking of some universal concept specifiable according to kinds. Between the plural of, say, warfare technologies and the singular of modern technology there exists a mere homonymy, the same word being used with diverse semantic functions. The specter of essentialism is thereby chased.

I wish to show some implications of the distinction thus suggested between technology as, on one hand, the common name for innumerable techniques and, on the other, the proper name for modern universal *mathesis*. A technique, the know-how that the Greeks called *technē*, is operative whenever humans make some goal-directed use of their hands. As to modern mastery over nature via a universal mathematical project, this is something different from know-how. It is one epochal project of truth, best described as 'technicity.' Because I was asked to discuss Heidegger, I first develop a few traits of technicity so understood. Then I suggest that the discourse of epochal configurations, or stamps—diachronic topologies, such as Foucault's—cannot be Heidegger's last word about truth in technicity. As stamping our world today, technicity reveals something that saves. It reveals a more originary truth: the tragic condition of being.

An Epochal Truth

Heidegger's description of modernity as the age of global technicity is probably well known. I therefore remain content here with enumer-

---

1. René Descartes, *Discourse on Method*, pt. 6, *Oeuvres de Descartes*, ed. Charles Adam and Paul Tannery (Paris: Vrin, 1965), 6:62.

ating a few thematic points—not Heideggerian theses but short cuts connecting some of what he called his 'ways' of thought. He has indeed followed diverse ways of thinking about *technē* and technicity, as well as about technology (in the sense of the *logos*, that is, the peculiar mode of gathering proper to technicity) and about technocracy (the regime imposed by technicity on thinking and acting). From his most explicit text[2] about these issues the following reminders can be spelled out.

An essentialist difference. "Technique is not the same as the essence of technique" (*QCT*, p. 4). Any modern procedure based on quantification requires a technique. But the essence of such techniques—'technicity'—is not itself technical. One may call this the technological difference, as quantification gathers *(legein)* phenomena into their way to be. The difference is clearly modeled, albeit for modern mathesis only, after the generic notion mentioned earlier. The essence *(Wesen)* of technique, Heidegger indeed says, is comparable to "the essence of a tree" (ibid.), which is not any particular tree. Likewise, technicity is not any particular technique, not this or that quantificatory procedure. The technological difference describes technicity in analogy with the subsumptive concept I began with.

A historical difference. "Enframing is destined to us as one mode of unconcealment" (*QCT*, p. 29). With this, modern technicity is now placed outside of generic descriptions. The displacement results from another sense of *Wesen*, other than Platonist: "Thus far we have understood 'essence' in the usual meaning it has [. . .] in the academic language of philosophy" (ibid.). The other sense of technicity, however, accessible to an 'other thinking'—other than subsumptive and opening, as we shall see, an 'other possibility'—inscribes our age within a concrete historical destiny. *Wesen* here is a verb. It designates the way epochs have unfolded in the West. In each, everyday stuff revealed itself situated within one finite constellation, one definite mode of unconcealment. Thus the crossbow does not belong in the age of technicity but in what Heidegger elsewhere calls the imperial-imperative epoch.[3] The stamp *(Prägung, Gepräge)* that marks our own age is described as *Gestell*, 'enframing' or 'positionality.' It differs from

2. Martin Heidegger, "The Question concerning Technology," in *The Question concerning Technology and Other Essays*, trans. W. Lovitt (New York: Harper, 1977), pp. 3–35, hereafter cited as *QCT*. I do not always follow this translation.

3. For the entire Latin-speaking age and beyond, the "fundamental relation to beings as such was administered by the sway of *imperium*" (Heidegger, *Parmenides* [Frankfurt: Klostermann, 1982], pp. 65–66).

things, not logically, as a species differs from the cases it encompasses, but phenomenologically, as one pervasive mode of manifestation differs from what it makes manifest. Enframing is not a universal. Nor is it a particular, in any way subsumed or subsumable under some universal. It is our *singular* site. By way of anticipation, let me add that tracing the epochal stamps by which the West has lived is one task incumbent upon topology.

Phenomenality. "Modern technicity [is] an unconcealing that requisitions" (*QCT*, p. 17). This answers the question of how, under the exclusive mode of manifestation that is technicity, phenomena are constitued. Modern phenomenality is due to a certain hegemonic trait in self-disclosure, which Heidegger calls 'requisitioning' *(Herausfordern)*. Phenomena appear, summoned as it were by the all-encompassing regime of method in the sense of Descartes.[4] Stated negatively, nothing can become manifest that does not conform to the project of global mastery. Nothing has being unless it enters into the system of quantification with its consequences, such as experimentation, production, circulation, consumption—in a word, unless it is objectified into "standing reserve" (*Bestand; QCT*, p. 17). This epochal requisitioning designates "the way in which everything comes to presence" (ibid.). It includes therefore the subject as well, which is why in the age of hegemonic technicity man becomes properly a phenomenon to himself only via the human and social sciences. Phenomenality no longer results from the intuitive grasp of an essence or species but from the probative grip on a specimen.

Truth as event. "In enframing happens the event of unconcealment" (*QCT*, p. 21). Throughout his writings, Heidegger sought to understand truth as a process of unveiling rather than a state of conformity. *Alétheia* is the originary condition beneath received normative notions of truth. Thus, when the norm is sought in 'adequate' propositions about data, or in the 'agreement' of a thing with its form, or again in some 'assimilation' of the soul to the true, the philosophical gaze remains riveted to things disclosed rather than moving back to the process of disclosure. With this step back from the manifest to manifestation, Heidegger steps out of the realm of metaphysics. As there have been various modes of manifestation in the West, truth has a history. The condition of that history is an event: truth always occurs as an unconcealing. This relation between condition and conditioned

4. See also Heidegger's commentary on Newton's First Law of Motion in *What Is a Thing?*, trans. W. B. Barton and V. Deutsch (New York: Regnery, 1967), pp. 76–80, 85–88.

in history was the one issue Heidegger kept pursuing in his middle period. It can be described as the difference between two disparate singulars: the unfolding of truth epochs and the ever-new truth event. Contemporary technicity is one such differential configuration, one epochal figure of manifestation as it differs from everything manifest. In it happens our epochal truth.

Conflictual Truth. "The essence of technicity is ambiguous in an elevated sense. Such ambiguity points into the mystery of all unconcealing, that is, of truth" (*QCT*, p. 33). Normative truth stands and falls with univocity. Indeed, received standards of conformity had to be unambiguous, simple. How else could they have normed performances of adequation, agreement, or assimilation? Understood as event of manifestation, on the other hand, truth is inescapably conflictual. A strategy toward concealment asserts itself in all modes of unconcealment. This ambiguity entails decisive consequences for the law as well as for being, affecting the very ultimacy of *Sein*. In terms of the law, a transgressive strategy works within every legislative strategy. In terms of the event that is being, an expropriating strategy *(Enteignis)* disperses all appropriating strategies *(Ereignis)* as they constitute constellations of phenomena. To render fully what Heidegger takes to be 'the Greek experience,' *alétheia* then needs to be understood as unconcealment-concealment; the law, as legislation-transgression; and being, as appropriation-expropriation. The originary duality indicated by the privative *a-* and the verb *léthein* (to dissemble) is in no way symmetrical. Despite their lexical parallelism, unconcealment is not the determinate negation of concealment. Unconcealment is of entities, but concealment is of being qua being. Truth and untruth therefore remain irreconcilable. The two strategies in truth, Heidegger adds, "draw past each other like the paths of two stars in the sidereal revolutions" (*QCT*, p. 33). With this disparateness at its very heart, truth loses its univocity. Again by way of anticipation: retrieving the event of disparate, conflictual truth will be the other task assigned to topology.

That which saves. "Where danger is, also grows that which saves" (*QCT*, pp. 28, 34, 42). In his investigations into technicity, Heidegger liked to quote this line from Hölderlin. What is that which saves? It can only consist in the express retrieval of truth's conflictual way to be. Hence this highly polemical either-or: *either* "pick up all standards" from among the unconcealed available stock, *or* "letting oneself be taken into the essential way to be of the unconcealed and of its unconcealment" (*QCT*, p. 26). This either-or sets "the other thinking" apart from all normative, subsumptive, entitative, univocal thinking. It sets 'the other possibility' apart from technicity as today's actuality. The other thinking and the other possibility indicate that which saves, namely, that man place himself expressly within the double bind that is truth. "As yet unexperienced, but perhaps more experienced in a time to come, [man] may be needed and used to preserve truth's essential way to be" (*QCT*, p. 33). Since that essential way to be is conflictual (from *fligere,* to 'clash'), polemical (from *polemos,* 'strife'), agonistic (from *agonia,* 'contest'), there is agony in salvation. In the words of the epigraph quoted at the beginning, that which saves resides in the possibility and necessity of tragedy.

FROM TOPOLOGY TO THE TRAGIC CONDITION OF BEING

This is obviously not the place to propose a theory of Attic tragedy. Suffice it to say that it always displays a clash, a strife, a contest between irreconcilable laws: for example, between those of Antigone and those of Creon.

Tragedy traces out something like a path of sight. The hero *sees* the laws of the city and of the family in conflict. Then he *blinds himself* to one of them, keeping his *gaze fixed* on the other. Nations and other communities have lived and continue to live within the shadow of this blindness. All normative single binds result from such self-incurred blinding. Then follows a catastrophe that *opens the hero's eyes.* This is the moment in which tragic truth is recognized. The vision of irreconcilable differing among laws *takes his sight away* (even gouges his eyes, as was the case for Oedipus and, in another way, for Tiresias), and it singularizes the hero to the point that the city no longer has room for him. From denial to recognition, blindness is transmuted. His orbits empty, Oedipus sees a normative double bind. He sees tragic differing. In *Oedipus at Colonus,* he is shown embracing it, reduced to silence but granted apotheosis, that is, deification, and receiving tutelage over Athens.

Heidegger's topological inquiry into the epochal stamps by which the West has lived, and of which technicity is the latest, models its

itinerary after that tragic crisis.[5] It looks back toward beginnings that
have initiated epochs and discovers in them an ambivalence breeding
hubris. Then it seeks to grasp—again in the words of the epigraph
above—the mutations in the essential way to be of truth. Heidegger
seeks to retrieve the critical 'thrusts of time,' as he also calls the
reversals in the history of truth, which set epochs apart. Finally, topology finds its own task in keeping alive the question of a possible
there, a possible topos in which humankind might prove experienced
enough to sustain the tragic condition of being.

That topos would respond to the early Greek. Here are a few lines
on the Greek beginning, as well as on the other beginning in which
tragic knowledge might save us in—not from—global technicity: "All
inceptions are in themselves what is complete, unsurpassable. They
escape historicist recording not because they are trans-temporal and
eternal, but because they are greater than eternity. As *the thrusts* of
time they place the openness for being to conceal in. The proper
founding of this time-space is called *being*-there" (*BzP*, p. 17).

Topology is an inquiry into topoi, places. In Heidegger, these are
the epochs in our history as they have led up to the age of technicity.
Topology shows the deep historical roots of today's global reach. As
it traces the places that have been ours, are ours, and can be ours, it
is recapitulatory, critical, and anticipatory.

Recapitulatory topology asks, What has been the form of conflictuality—of the tragic—that in its Greek, Latin, and modern ministry
the philosophical civil service[6] had both to see and to deny? In this
its retrodictive task, topology spells out the double bind marking
epochal inceptions. These are 'in themselves what is complete, unsurpassable.' Yet they are never simple. In his more nuanced readings of
the tradition, Heidegger tries to show how any *instituting* discourse
remains faithful to the phenomena, that is, in his words, to the dis-

5. That model is most notable in Heidegger, *Beiträge zur Philosophie* ("Gesamtausgabe,"
vol. 65 [Frankfurt: Klostermann, 1989]) cited hereafter as *BzP*. The present article is fully
understandable only in conjunction with three related papers on *BzP*: "Ultimate Double
Binds," *Graduate Faculty Philosophy Journal* 14 (Fall 1991), pp. 213–36; "A Brutal Awakening
to the Tragic Condition of Being," in *Art, Politics, Technology*, ed. Karsten Harries (New
Haven: Yale University Press, forthcoming); and "Riveted to a Monstrous Site," in *The Heidegger Case: On Philosophy and Politics*, ed. Joseph Margolis and Thomas Rockmore (Philadelphia:
Temple University Press, 1992), pp. 313–30.

6. "We are the functionaries of mankind" (Edmund Husserl, *The Crisis of the European
Sciences and Transcendental Philosophy*, trans. David Carr [Evanston: Northwestern University
Press, 1970], p. 17).

parity of unconcealedness and concealment. Epochal institutions 'place the openness for being to conceal in.' They point up the legislative-transgressive double bind, analogous in this to the opening of the dramas of Creon, Agamemnon, Eteocles, and others.

Any *instituted* discourse, on the other hand, represses dissension. As Heidegger diagnoses it, such repression is the phantasmic work of 'logic.' Following his less nuanced readings of the tradition, the reign of logic repeats the hubristic moment in tragedy when the heroic edict brings the conflict between city and family under one law. The order so established, however, never fails to prompt the return of the repressed. In the same way, even as it sustains an epoch, every hegemonic referent unfailingly also calls for singulars to assert themselves against subsumption, whereby it calls for its self-destitution.

In its recapitulatory work, topology inquires about the places—the phenomena, the regions of experience—that philosophers have retained to the exclusion of others and that they have absolutized so as to deny originary strife. Among these topoi have figured mathematics (Plato), motion (Aristotle), the observation of consciousness by itself (Descartes), and many more. From the elevated position of these strongholds, like Creon settled in his royalty, the normative *chargés d'affaires* have been able to legislate. They have exalted one representation beyond measure, hubristically declared it the standard of all measurement, and thus turned singulars into particulars subjected to it.

Critical topology. Here one topos only asks to be situated and described. It is that caesura place from which speaks the topologist of the present who asks: How can the double bind once glimpsed (the repressed 'having-been' of Greek tragedy) today appear imminent (the 'to-come' that saves and is a matter to safeguard)? In this literally decisive place, phenomenology does not, to be sure, turn into yet another instituting discourse. Rather, it addresses a transition. What is at stake is how "the essence of technique [. . .] can be led into the transmutation of its destiny" (*QCT*, p. 39). In that *krisis*, some problems disappear, while others, and formidable ones, emerge. All have to do with the status of a discourse trying to remain faithful to the singular that is happening to us and that is not simple.

What disappears is the very possibility of obliterating originary strife. The hypothesis of an epochal closure gives to the tragic double bind that singular feature, never before seen, of archic-anarchic monstrosity. Such bifrontal character of our site results from a 'thrust of time.' The phrase answers to the perplexity before singulars, in which historical thinking begins: things are one way, then suddenly in an entirely different way. In topological terms, we find ourselves inscribed within one horizon, one "time-space," and then suddenly reinscribed within another. This perplexity involves nothing but singulars: a *terra deserta* that is no longer ours, a *terra incognita* that is not yet ours, and a transition in which an entire collectivity lives as if holding its breath. No encompassing *nomos* governs this nomadism where allocations, assignments to discontinuous planes, follow one another without warning or mediation. Reinscription, however, institutes a possibility that remains to be seized upon. To *whom* can and ought the "proper founding" be entrusted? Not to any particular hero. The possible concerns precisely the way to be of the entire age of technicity. To the question, Who? the answer in Heidegger is a desubjectivized self, that is, a site (not a subject, which would answer instead to the question, What? What is man?). The transmutation of blindness about technicity, as described, cannot be forced. One can only let oneself be assigned to this conflictual site. Such an assignment may happen or not. It eludes willful conquering. When in a crisis, which may be ours, such allocation occurs, the originary condition of what we are living becomes thinkable. Univocal ultimates *(archai)* then appear as having covered up an an-archic originary condition; technical expertise proves to feed on the denial of tragic knowledge; global technicity reveals 'that which saves'; and in Heidegger's idiom of the mid-1930s, being-there *(Dasein)* comes to be. Since it is conflictual, the *there* of *Dasein* can only be tragic.

Yet the discourse of this historical crisis—critical topology—also encounters formidable difficulties. To trace the break between the univocally archic terrain and the tragically anarchic, it cannot but speak two languages. To think about the faultline *(de linea)* where epochal history terminates[7] requires *one* discourse uttered from within, as well as *another* uttered from without, that closure. It requires an epochal bilingualism. Thus we have yet to become, Heidegger says, "dwellers along the [abyss] in the time-space from which the gods have fled" *(BzP,* p. 52). The age of global technicity is readily described as one of great destructions, and not to subscribe to this diagnosis would take a hopeless case of anorexia toward reflection. But if to think is to stop at the conditions of what one is living, then

7. "The history of Being is at an end for thinking *in* the event of appropriation" (Heidegger, *On Time and Being,* trans. Joan Stambaugh [slightly modified] [New York: Harper & Row, 1972], p. 41).

it is worthwhile asking, How did we get here? To this question, the topology answers by pointing out the *krisis* around which the history of being turns, which enables us to become the dwellers along the abyss that we are. Critical topology implies that in the twentieth century the double bind beneath every monofocal posit becomes obvious, which in turn implies that the tragic denial at the foundation of order makes itself all the more violent. Institutionalized brutality of a magnitude never seen before, error and errancy, an isomorphism in everyday life that stifles reflection; then the critical turning that moves outside of the same and its laws; lastly the lesson learned from devastations, the vision of empty sockets and the silent affirmation of originary differing: this tragic itinerary becomes accessible only to a bifrontal phenomenological thinking. The topology that frees dissension is a thinking, since Heidegger does no more than linger upon the site and its conditions; it is phenomenological, since it gathers what shows itself there; and it is bifrontal, since it looks back at the collapse of normative epochs (strictly speaking, a redundancy: an epoch always results from an idea that puts a stop to questioning, from a fixed idea, a half posited, a decree, and, in that sense, from a norm) and looks ahead at a constellation that is anarchic because it *lacks simple ultimacy.* What shows itself is the kenosis that makes possible, while deferring it, the tragic *there* of being-there.

The difficulties have less to do with the circular character of this topology (which surmises a transition at work and then discovers symptoms verifying the surmise). Not every hermeneutical circle is vicious. The difficulties arise rather from the fact that topology can be retrospective and critical only because it is essentially anticipatory. In the vocabulary of representation, the promoting agency behind all standards, both the meta-narrative instructing us about the stage whence we come and the semi-narrative about the possible *exit* we are living (semi-, since narration remains suspended as philosophy runs out of representables) have their condition in a discontinuous sense in which all actors today find themselves enrolled.

As this is a temporal discontinuity, the preeminence of the possible over the actual[8] requires a "headway in order to reach our 'location' in being itself and thereby our history" (*BzP*, p. 501). We learn about our past and our present only via the contingent future, that is, via

---

8. Heidegger, *Being and Time*, trans. John Macquarrie and Edward Robinson (New York: Harper & Row, 1962), p. 63. The key section on the mathematical project as well as on the genesis of the theoretical gaze is sec. 69, trans. pp. 401–18.

the eventual, eventlike topos. As long as the tragic *there* has not been seized upon as our ownmost possibility, one will never "measure what has happened in the history of metaphysics: the prelude to the event itself" (*BzP*, p. 174). This prelude has from all time deferred the tragic *there*, as well as the differing whose possible site it is.

Heidegger is not saying that soon, perhaps in the year 2000, we shall reach our conflictual location in being and thus be saved from the global reach of technicity. He is saying that the tragic *there* is possible, that it always has been, that it has become even more clearly so in the age of worldwide cataclysms that is our own, and that there is no bearer of salvation (*salvus*, as 'whole' as tragic vision becomes after the crisis) other than that possible. With the deferred *there*, nothing gets postponed until later. Deferment, rather, describes originary differing itself.

Deferred in what way then? The question is one of anticipation.

Anticipatory topology. This deals with a possible historical locus: one already given, yet still to be occupied. Its description is first of all negative since topos here no longer signifies any region of entities whose relations can be maximized to produce some archic referent. "In the other beginning, it is no longer either some entity or some region and realm, any more than entities as such, that can set the standard for being" (*BzP*, p. 248). The difficulty consists in understanding anticipation without associating any utopian or millennial postponements. This requires some clarification of the negations in the lines just quoted.

Clearly what is being denied under their obvious apophatism is the hyperbolic use of one entity—proper (doubtless the Platonic 'good'), regional (doubtless Aristotelian 'physical' motion), or general (doubtless Leibniz's perplexity about why there are 'entities in general')—posited as normative. They are aimed against representational maximization. What is excluded is the transition to another genus as Hegel had described it. The "great men," Hegel said, "see the very truth of their age and their world, the next genus, so to speak, which is already formed in the womb of time."[9]

But what is thereby being denied more broadly is a certain phenomenology of simple manifestation, the kind one may be tempted to read into Heideggerian 'giving.' Such a phenomenology, answering the ancient question of being qua being, would run like this:

---

9. G. W. F. Hegel, *Reason in History: A General Introduction to the Philosophy of History*, trans. R. S. Hartman (New York: Liberal Arts Press, 1953), p. 40.

Heidegger relates *geben* to the Greek *phain-* and *phy-*, the radicals for 'showing' and 'rising,' hence to manifestation; then he inverts the relations of grounding (as attested to by the lines quoted), anchoring not manifestation in one entity that shows itself most noteworthy but rather this entity as well as all others in manifestation. Being qua being would consist in simple giving *(Es gibt)* as underscored by these verbs in the middle voice: 'to show forth,' 'to rise.' To respond to the question of being, presencing would have to be retained as such, regardless of any consideration for present entities.[10]

This is all well and good. Still, reconstructions such as this say nothing about the *intrinsic contrariety* of being, nothing about the originary strife for the sake of which Heidegger never ceased renewing his idiom. Being is indeed showing, rising, manifesting, presencing, giving; but it is all that in a 'fissured' mode, in 'dissension,' as 'unconcealment-concealment,' 'appropriation-expropriation,' in the 'struggle' by which the No asserts itself against Yes and death declares itself against life. Whatever apophatism there remains in Heidegger, it does not hide some negative haplology *(haploon,* 'simple').[11] It puts simplicity out of operation. With their negations, the lines quoted do suggest that being alone is *maßgebend,* setting the standard. But "Yes and No are an essential property of being itself, and the No even more originally than the Yes" *(BzP,* p. 178). With essential properties like these, how would apophatic utterances ever exalt a simple standard, retrieve a simple condition beneath epochal regimes, prepare a simple dwelling site? The *thought* of the double bind that is being may indeed be simple *(einfach, schlicht,* etc.), just as the sight of Oedipus blinded alone could be simple, whole, safe.[12] But if, in the other beginning, it is no longer any entity that ultimately 'gives'—gives (sets) the standard, gives (provides food for) thought, gives (brings about) a site—nor any thesis about being that binds uniformly, then Heidegger's great awakening after the compromises of 1933–34 led him outside of all *simply normative* regimes. Anticipa-

10. Heidegger goes rather far in formulating this disregard: a "knowing seriousness" would "no longer be bothered by good and bad, by decay and rescue of the tradition, by benevolence and brutality" *(BzP,* p. 242). See also *On Time and Being,* pp. 2 and 24.

11. For one recent attempt to read into Heidegger just such a negative haplology, see Jacques Derrida, *Of Spirit: Heidegger and The Question,* trans. G. Bennington and R. Bowlby (Chicago: University of Chicago Press, 1989), pp. 108–13; and *Psyché* (Paris: Galilée, 1987), pp. 535f.

12. Oedipus blinded "has one eye too many perhaps. . . . To live is death, and death too is a life" (Friedrich Hölderlin, "In Lovely Blueness," in *Poems and Fragments,* trans. M. Hamburger [modified] [Ann Arbor: University of Michigan Press, 1967], pp. 603f).

tory topology leads toward a conflictual normativity, the knowledge of which is more ancient in the West than any theticism of the same—the theticism most effectively deployed with contemporary global technicity.

To speak of the deferred tragic *there* is to sustain the possibility that the denial of the transgressive other in what philosophers ascertain as ultimate may reveal itself exhausted. Such denial has provided the impetus to every figure of the legislative same, while its exhaustion has always secretly shattered normative power. To attempt the possible means here to destabilize affirmation, the firm, firmness, and to rehabilitate under the Yes the pull of the No. The possible undercuts the effects of positive, that is, thetic discourse. Hence the silence of the tragic hero fleeing into that *nonplace* for the ancients, the desert; a nonplace *(ou-topos)*, not moving (like the island of some messianic society) into an ever-greater distance as a result of a poorly calculated arrival, but deferred under the mode of the possible and therefore differing with itself. No utopia here, and quite another apophasis. Negating and questioning are all that remain to anyone seeking to retain the disparate: for Heidegger, a negating that posits no dialectical nescience and a questioning that presupposes no maieutic prescience. The topology of the possible does not work with a theodolite. Like the desert where Oedipus wandered, the anticipated conflictual topos is closed to prospection. In the *Beiträge*, it is described through 'fissuring' *(Zerklüftung)*, a word that stresses time more than space. To be deferred is the very temporality of the tragic *there:* the discord of phenomenal presencing and of a singularization always to come (see below). This is the originary strife whose vision blinds one and whose knowledge strikes one silent.

Now if such is the originary condition as Heidegger has come to view it after his brutal awakening from reliance upon univocal "originary rootedness,"[13] then why state it in the future tense? Witness the proliferation of topical postponements: at present, we can at most prepare the "place of questioning" *(BzP*, p. 85), where "those to come" will prepare themselves, who in turn will prepare "new standpoints in being," from which alone can "the strife of earth and world be sustained" *(BzP*, p. 62), a strife out of which may issue the tragic *there.* This series of sites—*Frageort, Standort, Da*—has nothing to do with a speculation on happier morrows. In anticipatory topology, the

13. *Ursprüngliche Verwurzelung* (not *"primordial roots,"* as the translators put it), *Being and Time*, p. 429.

emphasis remains on the possible site and its discontinuity rather than on actual terrains and their contiguity. Statements in the future tense concern a potential that is given now. They are not meant to put off indefinitely the other beginning, but to aggravate here and now the discontinuity between regimes of the same, whose institutionalized hypertrophia Heidegger had just helped in promoting, and the event of differing by which he henceforth understands being—to aggravate, in his words, the temporal difference between illusory full presence and fractured presencing.

Providing a Place for the Originary Double Bind

If topology is concerned with the place where the conflictual event can be sustained, at least one misunderstanding regarding Heidegger is excluded. One might indeed object: if he advocates express compliance with an originary strife modeled after Aeschylus and Sophocles, then what is there about his model that *saves*? Is such a possibility not more disruptive than even global subsumption under univocal standards?

The objection is meaningless. It reduces the guiding category of modality, the possible, to what derives from it, the actual. One worries about what will happen to public order *if* the legitimating referents were to turn out fractured. This amounts to misconstruing the status of the possible, since conflictual being already grips and breaks us, just as it has always gripped and broken us: "We are already moving, although still only transitionally, within another truth" (*BzP*, p. 18): within the other truth which, in the epigraph quoted at the beginning, was described as the very condition of tragedy. In hardening the reign of rule-governed technicity, we would merely be once again answering the possible with denial. Planetary totalitarianism— Creon's worldwide and ultimate triumph—would only extend and deepen the actual. Creon forces and enforces the public order by denying its transgressive other; but what would there even be to deny if the tragic double bind did not hold him always and already? The conflictual possible is *to come* only because it is our most ancient condition. It breaks with the actual as the event breaks with constant presence; as the singular, with subsumption; as originary strife, with any one focal meaning of being; and as the acknowledgment of what is (letting the tragic *there* be) breaks with thetic repression (positing phantasmic figures of firstness).

In order to see in what sense the originary double bind "saves" in technicity, it suffices to ask, As opposed to all thetic ultimates, what is phenomenologically ultimate in everydayness? Ultimacy is the property of a condition that cannot be sidestepped or stepped behind. A condition is ultimate if it is not determined by some yet more originative condition. In everydayness, only our coming-into-being and out ceasing-to-be satisfy that requirement. What is phenomenologically ultimate are not precisely birth and death as two biographical facts, one having occurred and the other yet to occur, but two pervading traits—two pulls on the present—originating in them. Hannah Arendt called the first of these traits 'natality'; it is "the new beginning inherent in birth," "the principle of beginning" "in which the faculty of action is ontologically rooted." Primal institutions are good examples of that originative trait, but so are drafting parliamentary legislation, founding the United Nations, sitting down to begin a book, or choosing a life partner. Arendt called the second trait, our being-toward-death, 'mortality.' It is what ties us to one definite sequence of events, bound eventually to lead to our extinction. Mortality is therefore a description of our utter singularization to come. Natality and mortality are what we all know, yet poorly. Since they are given only mediately, there two traits are what stand most in need of rigorous clarification. As phenomenological ultimates, they, more than any immediately given phenomenon, call for *diasozein ta phaino-mena* ("preserving that which shows itself," Eudoxus of Cnidus).

The double bind of natality-mortality is phenomenologically first in the sense of the *origin* of each and every experience. It is not first in the sense of *principles,* not foundational in any way. Principles and affiliated representations arise only from the subsumptive violence that turns whatever is the case into a case of whatever universal. Global technicity is the most effective translation of principial thinking into everydayness.

Now whenever philosophers ask, In the name of what? they refer facts and deeds to principles, which bear common names. Heidegger, for his part, seeks to step back from principial referents to what is originary and does not bear a common name. Natality and mortality, as mine, are *singular* traits. Topology is thus a discourse about singulars not only as it narrates the sequence of epochs in our history but more decisively as it seeks their condition in the originary double

bind of natality and mortality. The one great possibility never to become an actual economy—which is why there is no salvation *from*
technicity—would be to comply equally with the claims of singulars
and the phantasms of the common: to comply with the temporal
difference. Topology does not legitimize facts by referring them to
a name. It legitimizes the feasible in anticipating the one nameless
possibility, the event of appropriation-expropriation. What common
name indeed would fit differing laws? In Heidegger, the whole issue
of legitimation gets transmuted. The site of norms is to become what
it has always been: the locus of the event, a locus to which we have
always been allocated already and whose evidence gains as principles
collapse into the isomorphic: "Being happens as the event. This implies that a singular site has unexpectedly befallen us, alienating us
toward the instant and only thus spreading" (*BzP*, p. 260).

It is true that the locus thus anticipated can no longer guarantee
any tranquillity of order. Yet this is not to say that its topology justifies
anything and everything. Its criteria of responsibility were set forth
not only by Aeschylus and Sophocles but also by Heraclitus and
Parmenides. It heeds other imperatives than do normative signifiers:
"The things that are absent, behold them nevertheless as firmly present."[14] In the thetic impulse of natality behold the undertow of mortality. In the site mapped out by the law behold the singularization
to come, which no law can subsume and which alienates us from the
lures of normative presence.

"Must we then not think otherwise, anticipate zones and yardsticks
and modes of being that are entirely other, so as to belong to the
necessities that are dawning here?" (*BzP*, p. 204). Such is the onus of
anticipatory topology. It is a matter of belonging to the originary
possible, more powerful than any actual denial. But in this possible,
a 'not yet' stands out from the 'already.' The instrumental nature
of normative single binds has become obvious at least since Marx,
Nietzsche, and Freud, but there is 'not yet' a readiness to acknowledge tragic double binds. Hence: "The locus of decision still remains
to be founded" (*BzP*, p. 187). But also: "To think this way already
requires, to be sure, a standpoint where it will no longer be possible
to be lured by all the 'good' and the 'progressive' and the 'gigantic'
that are being achieved" (*BzP*, p. 140). In what passes for good, progressive, gigantic—which probably means: in the technological gigan-

14. Parmenides, in *Die Fragmente der Vorsokratiker*, 6th ed., ed. Hermann Diels and Walter
Kranz (Berlin: Weidmann, 1951), frag. 4,1.

ticism that passes for progressive and hence for good—arise
necessities that compel one to think otherwise. The everydayness
that breaks with the cluster 'good' (again, Platonic), 'progress' (early
modern), 'giganticism' (late modern) is already ours, already a necessity in the sense that everyone is living it, but it still remains for us
to seize. The possible always grafts some *not yet* on an *already*, thereby
breaking the order of proper temporal sequence. To say a cataclysm
is possible is to say that at present things are holding up, but that
the conditions of their collapse are already there. Heidegger: the conditions—the zones and the yardsticks and the modes of being—are
there for the illusion to collapse due to which values, progress, and
technicity appear to us as pregnant with normativity. Insofar as they
remain still to be grasped, these conditions are situated ahead of us
(ante), whence anticipation. They fissure the subjectivist topos that
continues to pass for rock bottom, whence topology.

To establish oneself expressly on the fissured ground is what in the
mid-1930s Heidegger understands by 'being-there.' Such establishing
marks the other beginning. Its implications appear from a disjunction—a twofold disjunction—that Heidegger was to attempt ever
anew over the forty years to follow.

On the one hand, the *fundamentum concussum* remains to be disjoined from any foundation seeming *inconcussum*. That disjunction
devolves on a recapitulation, or an archaeology, which frees plateaus
in collision from under the numerous foundations posed since the
Greeks; which frees, he says in some texts, the strife between world
and earth. On the *world* map of phantasmic dominions he traces out
especially the modern enclosure, freeing under the discourse established since Descartes and Kant the *earth*bound dissensions that have
torn each of the great founding discourses. This labor of archaeological disjunction is incumbent on historical critique. It has its site and
its hour: the 'thrust of time' making us, the subjects of technicity, feel
the collision of plateaus that are presupposed beneath every foundation posed.

On the other hand, the forces in collision themselves remain to be
disjoined. Heideggerian responsibility consists in a response to that
destabilizing thrust in which normative posits suffer their deposal.
The 'thrust of time' brings into light the disparate beneath every
referential construct. To respond to what is happening to us then
means to lay bare forms of the originary double bind: appropriation-

expropriation in the event; the strife world-earth; the conflict gods-humans in history; that of unconcealment-concealment in truth; the discord of phenomenalizing-singularizing in manifestation . . . These polemological disjunctions fall upon anticipatory critique. It separates the dislocating strategies so that being-there can belong to its originary locus, that is, (Heraclitean) *polemos.*

The self-incurred stultification in technicity shows just how far away we still are from responsibility belonging to the originary locus so understood. Whence the appeal to "found the historical place for the history to come" (*BzP*, p. 60), to "create that space-time, the site for the essential instants" (*BzP*, p. 98). Later Heidegger was to describe the instant that founds history in an abbreviated form as the "entry into the event"[15] (abbreviated, since the phrase does not say the event results from mutually hostile pulls). This entry would be as compact as all beginnings have been. What interests the topologist in this compactness is the conflictuality glimpsed, then infallibly repressed, at the inception of every single epoch in our past.

Stultification gets squared in the consumer of ideas who has seen and read everything. There may be something charming about a blasé gymnasium student; later, it is better to have learned how to read. For example, on truth and its site. "The truth of being [has still] to find a site that comes to be in being-there" (*BzP*, p. 90). Having leafed through Heidegger, one may vaguely recall that, at least when read in translation, he liked to speak of authenticity; that this doubtless has something to do with truth; and that there were decades when fashion required that one 'be true.' Speed reading may, however, make one miss the essential—here, the conflictuality without sublation which is truth. Unconcealment-concealment has yet to find its site. 'Being true' then means something more disquieting than genuine or veridical existence, as it may seem at first reading. It means the condition for which both Hölderlin and Nietzsche "had to leave prematurely the clear of their days" (*BzP*, p. 204). The site of truth is "the there in its abyssality" (*BzP*, p. 240). Thus the conflictual *provides the place* for the abyssal.

Such places are imparted rarely. It is one thing to describe the conflict that broke tragic heroes first, Hölderlin and Nietzsche later. It is another thing entirely for an age to find itself allotted the double bind. Topology carries out the description; the event, allocation.

15. Heidegger, *On Time and Being,* p. 41 (translation modified).

Providing a place is then not primarily a human achievement. Here again, misconceptions lie in wait for the reader of Heidegger. One of these has to do with the desire for favor. To 'provide a place' (*verstatten*, e.g., *BzP*, p. 278) means, to be sure, to impart, assign, permit, grant. But *to what* does the event accord a place? If it is to unconcealment-concealment, and if the event is itself appropriation-expropriation, then our site (*Stätte*) is never accorded to us in a mode other than that of discord. To allow oneself to be charmed by the Heideggerian 'there is' and 'favor' amounts again to reading poorly. Each place is allotted with grace, but never without disgrace either.

Another misconception has to do with the desire for a solid ground. Heideggerian nostalgia would want to attach the people to its soil, as the Athenians were attached to Attica. And, indeed, does he not celebrate rural rootedness? Yet the history of being knows no privilege: no *lex priva*, no law that is private, tnat is, deprived of its transgressive other. The 'there is' occurs always, and always the giving is taken back as well. Always the *Es gibt* sunders one's dwelling site. Metaphysics was born from a desire wishing away allocation-dislocation as unbearable. Since the classical Greeks, denial has intervened normally, equipped with the apparatus of norms and the normal. In Aeschylus and Sophocles, undoing it took an act of parricide or fratricide. To us, it seems, even a world of technicity on the way to self-destruction does not suffice to undo tragic denial.

These two misconceptions suggest the difficulty of the topology. It is again a matter of desire, as Heidegger's 'other thinking' is being challenged to succeed where theticism could only fail. Having heard of metaphysical closure, one expects an investigation of contiguous historical spheres; and, remembering his exhortations to keep oneself ready for a 'gift,' one expects the demonstration of an entry into the good graces of fate. But Heidegger describes laws that always fracture the unconditioned. Gift, granting, grace, 'that which saves,' and so forth are thus to be thought of other than as healing. The thrusts of time bring no therapy to the truth of being, which is tragic strife.

Remaining riveted to a monstrous site (in accordance with the very hypothesis of closure), one expects at bottom a 'philosophy of the one' capable of edifying. Heidegger shows, however, that—and how—edification has proved hubristic and the floor it presupposed anything but solid. This is, as it were, the advantage of our epochal monstrosity: coming from afar, taken already by the polemical truth of being, how can we persist in the tragic denial from which were born all normative posits, including that of global technicity? Just like any other force of repression, the force of denial that sustains theticism works destruction. Its illusory edification is built upon the ruins of singularity.

What remains to be seen now is how the earth can be thought of other than as standing reserve for technicity; how, in other words, a singularizing pull always breaks up any pull of phenomenal contextualization into a world. "Why does the earth remain silent in this destruction? Because it is being denied the strife with a world as no place is provided for the truth of being" (*BzP*, p. 277f.). (Heidegger's suggestion in these lines that the earth's destruction would become utterable only with the recognition of originary strife should guard one from co-opting him for some ecological cause.)

The Tragic Condition of Being:
Contextualization-Decontextualization

> What is, when the struggle for standards is dying out?
> —*BzP* 28

If technicity, today's epochal stamp, obliterates originary strife even as it hardens it to the extreme, then 'providing a place for the truth of being' does not amount to as obscure a program as it may sound. Nor does it amount to that all too familiar program, dialectics: as if, speaking of what saves, Heidegger were expecting that positioning (or enframing, *Gestell*), when pushed to its extreme, would prompt its own negation. He is not claiming that the oppositional logic of the world spirit will grant some remedial antidote to theticism. In the mechanics of determinate negation, theticism triumphs.

What is being asked of our age for the sake of 'salvation,' it seems, is to *undeny* a twofold pull in whose grips technicity places us more brutally than has any other epoch. Technicity contextualizes things and humans under the most efficacious universal ever seen, yet it also singularizes them without world or context. Such is the conflictual condition by which it functions. Knowing that condition is what saves. Technicity can instruct us in being's law,[16] which is ultimate

16. "In enframing we glimpse a first, oppressing flash of the event" (Heidegger, *Identity and Difference*, trans. J. Stambaugh [slightly modified] [New York: Harper & Row, 1969], p. 38). We glimpse the *event*, namely, of appropriation-expropriation; we glimpse it in a *flash*, since technicity pushes both subsumptive universalization and dispersive singularization to

without being simple. It can teach us that to rely on univocal ultimates is to cover up contextualizing-decontextualizing. These two strategies account for being qua event. How do they function?

Isomorphic contextualizing. If the task of the philosophers has been to ensure subsumption of all possible phenomena under one universal thesis of being, then today we have all become functionaries in Husserl's sense: all experts, that is, in treating whatever is the case as one more case to be subsumed under the modern primal institution (*Urstiftung,* which for Heidegger always results from a 'thrust of time' establishing a new focal sense of being). Such is the scientistic a priori by which we live. As Heidegger views it, modern science is never a disinterested inquiry. It is the tool for a highly interested posture we cannot avoid epochally in all our moves. Turning outward, we find but natural resources to be harnassed, and turning back upon our past, only cultural curiosities to be marketed. Turning inward, I no longer find, as did Augustine, the immutable light of truth, but rather neuroses awaiting treatment; and turning upward, more resources—perhaps uranium on Mars—awaiting exploitation.

As an epochal project, contextualization through technical-scientific mastery leaves no residue. When that project instituted the modern age, why indeed did substances facing the subject need to be termed 'extended,' if not to make them all equal *(iso)* in their forms *(morphé),* that is, uniform, isomorphic? According to Kant, the uniformization of phenomena is obtained through the inner sense. This means that only what can be temporalized is a datum and that only what can be ordered in calculable succession or simultaneity is temporal. As for the acts of the understanding, there are powerful reasons why the first of these acts must concern what can be measured and calculated. Since the subject knows only perceptions that are measurable in their spatial and temporal extendedness, the first Kantian category has to be that of quantity. Thus the uniform and the quantifiable must not be considered more consequences of the modern 'Copernican revolution' toward the legislating subject. Quite the reverse: only because every phenomenon has to be accessible to calculation—because its phenomenality is summed up in its calculability—does the subject have to occupy the place of central legislator. Establishing itself as the spontaneous source of the laws of being, the subject will be certain that these laws are the same everywhere.

their extremes; that flash is *oppressing,* as technical universalization institutes global violence; yet the glimpse is a *first,* hence there is salvation in global technicity.

The modern regularity of phenomena results, then, from an act of anticipation, from an a priori positing, a thesis. For Heidegger, the spontaneous subject is defined by this very theticism alone. *"Positing a rule ahead"* (*BzP*, p. 161) is the one posture that constitutes both the autonomous agent and the datum uniformed by him; both the self and its other. Now a rule prescribing what ought to be is called a 'norm.' The isomorphic is the *normal*. To the Baconian and Cartesian institution of the project of normalization responds its full-blown success in the twentieth century. To Heidegger's question "What is, when the struggle for standards is dying out?" a first answer is obvious: struggles, for example, ideological ones, belong to the past, as there remains only one standard, contested neither East nor West—universal isomorphism. This is what being-in-the-world has come to mean in our age. The contextualization of phenomena, that is, the way they become phenomenalized, is everywhere the same.

Dispersive decontextualization. Yet as that every mode of phenomenalization, technicity also singularizes them without any agent of reconciliation. Be they phenomena of nature or of culture, subjective or objective, encountered outside or inside, they have lost their respective worlds. Natural resources, neuroses, archaeological or literary curiosities *are* inasmuch as they answer the question: What can be made with this? And making, *poiesis*, is always of singulars.

The point is decisive. As long as one does not see the enduring strategy of *making* at the heart of the age-old mechanics of subsumption—at the heart, therefore, of 'logic'—the counterstrategy of singularization is bound to remain incomprehensible, and with it the originary double bind that Heidegger seeks to rehabilitate.

This can be shown with regard to the ancient notion of *phusis*. Just as the English 'nature,' Greek *phusis* means two things: on one hand, a certain region of phenomena and, on the other, a principle of production. It designates the region of things not made by human hands, and it is the principle by which the things belonging in that region—plants and animals—are produced. The semantic ambiguity can be illustrated by a nasty two-sentence joke that circulates in London about a member of the royal family: "Princess Anne likes nature very much. That is astonishing if you consider what nature has done to her." She likes nature, namely, that region of phenomena where she finds trees, meadows, and horses, not buildings, streets, and cars. As to what nature has done to her, she, like every other biped, is also a product of that principle of production by which trees, meadows, and horses grow, a cause whose products can be more or less attractive. In Aristotelian terms, *phusis* denotes both a kind and a cause: the kind of things other than artifacts and the cause other than art *(technē)*.

For Heidegger, this ambiguity shows that being has always been understood in terms of making, producing, causing. *Phusis* has all along been conceived according to its polar opposite, *technē*: first rather covertly, designating the sum of "those beings that produce themselves out of themselves," later overtly, *natura* designating beings in their totality as "the created" (*BzP*, p. 126). Throughout our history 'to be' has meant, in one way or another, 'to be caused.' By what? By such ultimate representations as "the Suprasensory World, the Ideas, God, the Moral Law, the Authority of Reason, Progress, the Happiness of the Greatest Number, Culture, Civilization," all standards qua causes. Heidegger adds that in the age of technicity these representations have "lost their constructive force and become nothing" (*QCT*, p. 65). At the moment when making remains the sole standard, all standards *for* making whither. The global reach so unfettered is what Heidegger calls "operational machination" (*Machenschaft, BzP*, p. 126).

To the question, "What is, when the struggle for standards is dying out?" the answer now is: there are only singulars, as each object for operational machination sets its own procedural standard. In the mid-1930s that assertion of singularity was understood by Heidegger as having brought about 'the highest danger,' precisely operational machination in America, the Soviet Union, as well as Nazi Germany, and at the same time 'that which saves,' namely, the knowledge of the originary differing of contextualization-decontextualization. "Being—the remarkable erroneous belief that being should always 'be.' . . . Being is what is most rare because most singular, and no one guesses the few instants in which it founds a site for itself and unfolds" (*BzP*, p. 255). Being is ultimate, yet singular and rare, as the event of contextualization-decontextualization   (Heidegger   says 'appropriation-expropriation,' *Ereignis-Enteignis*) is not simple. Its site is the tragic *there*. Hence this other question: "Entering into being-there, its instant and its place: how does this occur in Greek tragedy?" (*BzP*, p. 374).

The double bind that saves. Heidegger neither deplores nor applauds the loss of enduring standards in global technicity. He reads that loss with the eyes of a symptomologist, and it reveals to him the self-assertion of the singular within, against, and at the basis of all normative posits. The analysis of technicity thus allowed him to translate into the ultimate condition of being what had earlier appeared as *man*'s condition: the double bind of natality-mortality.

To the question, "What is, when the struggle for standards is dying out?" he gives this answer: in every normative posit the strife of singularization against universalization *is* the standard.

Suggestions for Further Reading

Heidegger, Martin. *The End of Philosophy*, trans. J. Stambaugh. New York: Harper & Row, 1973.

Pöggeler, Otto. *Martin Heidegger's Path of Thinking*, trans. D. Magurshak and S. Barber. Atlantic Heights, N.J.: Humanities Press, 1987.

Schürmann, Reiner. *Heidegger on Being and Acting: From Principles to Anarchy.* Bloomington: Indiana University Press, 1987.

9 Postmodernity, Technology, Ontology

GIANNI VATTIMO

"Sociological Impressionism"

The thesis I propose concerning postmodernity and technology is not simply that there is a "postmodern" view of technology or, conversely, that there is a technological determination of postmodernity. This much is certainly true, but beyond it, and from a philosophical point of view, I shall argue that what is going on, what is happening to us, and what *concerns* us in the postmodern age is a transformation of (the notion of) Being as such—and technology, properly conceived, is the key to that transformation.

In my view, it makes sense to speak of postmodernity, and of the meaning of technology within it, only from a point of view, guided by the teachings of Martin Heidegger, that tries to develop what one could call an "ontologie de l'actualité" (in a phrase of Michel Foucault, where *actualité* means "the current epoch," but also "everyday life").[1] I propose to use the expression in a very literal sense: it does not simply denote a philosophy oriented toward the consideration of history and existence rather than of epistemology and logic. What I suggest we must develop under this name is, roughly speaking, a

1. In one of his late essays, Foucault opposes philosophy as *ontologie de l'actualité* to philosophy as *analytique de la verité*. Here I am using the expression in a completely independent sense, unrelated to Foucault's meaning.

214

theory (or, more simply, a discourse) of what Being means in our current situation. I am aware of the difficulties and dangers involved not only in speaking of "Being" but also in the effort to describe "our current situation." Indeed, the latter expression is even more problematic than the term "Being" itself. I would also suggest that the difficulties involved in this term are related to the fact that, since its very beginning, philosophy has tried to clarify "Being" by eliminating its relation to the other term under consideration here, the "current situation" of the speakers. Heidegger would say that it is for this precise reason that metaphysics has developed as a forgetfulness of Being, *Seinsvergessenheit*. From the Heideggerian point of view (which I largely adopt here, although in a specific and perhaps controversial interpretation), one of the first steps in going beyond the metaphysical *Seinsvergessenheit* toward a recollection of Being would be to try to take into account precisely this problematic side of the question, the *acutalité*, the "current situation" of our existence.

Furthermore, when one undertakes to examine the "current situation," one finds that its most salient characteristic is precisely a growing awareness of the need for this very turn to actuality. Much, indeed the greatest part, of twentieth-century philosophy looks like a sort of "sociological impressionism" (to borrow the title of Georg Lukács's study of Simmel). Think of Walter Benjamin, whose main unfinished work is a study of Paris as the capital of the nineteenth century; of Theodor Adorno, of Ernst Bloch, also of Jürgen Habermas, whose theory of communicative action, despite its "transcendental" appearance, is ultimately a theory of modernization; and, on the phenomenological side, think of Edmund Husserl's *Krisis*. The widespread philosophical concern with modern technology, which will soon become our theme, also belongs under this heading. But the greatest example of a philosophical sociology, or more specifically of an *ontologie de l'actualité*, is found in the whole of the late work of Heidegger, after the "Turning" *(Kehre)* in the 1930s, which followed the interruption of *Being and Time*. In Heidegger, the "sociological impressionism" that permeates much of the philosophy of this century finds a more precise and explicit formulation, one not to be found in the other thinkers who, in fact, practice this type of philosophy.

It is true that in thinkers like Adorno and Bloch the sociological attitude of philosophy seems to be justified from a general Hegelian-Marxist point of view. But neither Bloch nor Adorno adopts an orthodox dialectical philosophy—that is, a metaphysical view of history—that would be necessary to justify the tendentious "identification" of philosophy with sociology. Adorno, for example, takes his observation of the dialectics of enlightenment—the fact that present society realizes the "totalization" and integration that Hegel considered the final stage of reason and spirit—as evidence that Hegel and Marx were wrong because "totality is the false." Now, only from the dialectical point of view of the totality may philosophy be "identified" with the analysis of *actualité* (at least so long as one remains within the Hegelian-Marxist tradition, which Adorno never completely abandons). This contradiction between the rejection of a strong metaphysics of history and the "identification" of philosophy with sociology is even more visible in Benjamin, for whom the philosophical interest in the contemporaneous social reality seems to be justified from a point of view that, rather inconsistently, combines a still metaphysical view of history (of a Marxian kind) with a Platonic theory of the ideas (think of the "Gnoseological Preface" to the German *Trauerspiels*)[2] and also includes a "symbolistic" attitude reminiscent of Simmel (for whom a marginal social phenomenon like fashion *reveals* meaningful truths), as well as the Jewish tension toward messianic salvation. To take more recent examples, even J. F. Lyotard's thesis regarding the end of the "grands récits"—the global interpretations of history that form the basis of the dominating ideologies of our century—is in its turn (perhaps unconsciously) based upon a *grand récit*, since it is the postmodern transformation of the condition of our existence that has stripped those *récits* of their meaning and credibility. The same may also be true, in many respects, of Richard Rorty's idea of postmetaphysical culture.

What I want to suggest by these sketchy allusions is that the pervasive "sociologism" of contemporary philosophy is not fully self-conscious and therefore not "thematized" and justified in its implications. The only exception, in my view, is Heidegger, who explicitly speaks of an "epochal" essence of Being. After the "Turning" *(Kehre)*, which followed *Being and Time*, the throwness in which man finds himself always-already—which, in that first work, might still be taken as a sort of Kantian a priori—becomes explicitly *historical*, qualified from time to time like a natural-historical language (having no Chom-

2. Walter Benjamin, *Ursprung des deutschen Trauerspiels* (Frankfurt am Main: Suhrkamp, 1965). In English: *The Origin of German Tragic Drama*, trans. John Osborne (London: NLB, 1977).

skian ahistorical structure). Since Being is not to be conceived as an objective given that is prior to the application of all "conceptual schemes" (because all supposed precategorical objectivity is already based upon a conceptual scheme, as, for example, modern scientism), one may speak of Being only at the level of the historical events in which the temporally varying ways of the constitution of the world, as experienced by a historical humankind, actually happen. Being is not the object but rather the "disclosure" within which subject and object, man and world, find the possibility of their relation. As the disclosure cannot be attributed to the stability of the object, which can be given only within a specific disclosure, Being is event: Being *is* not, says Heidegger, it "happens" *(es gibt)*. It is only from this perspective, I believe, that it truly makes sense to "identify" philosophy with sociology or with *ontologie de l'actualité*. The recollection of Being, overcoming the metaphysical forgetfulness of it, which Heidegger wants to prepare through his work, is an effort of remembering the *Ge-Schick* (destining)—the connections *(Ge)* of the sendings *(Schicken)* or of the disclosures that have determined and made possible the experience of the historical humankinds. Only by inserting our present *Schickung* (sending)—which I would translate as the current meaning of Being—within this connection can we recall Being and not simply identify it with "the beings," with the present order of existence.

I have given this very rough summary of what Heidegger means by the epochal essence of Being in order to explain why and in what sense philosophy must turn to *actualité*, must speak of our present situation, of modernity and postmodernity. It must do so not as the application of some more general system to a specific case—our society—but rather in order to grasp what Being is in the current disclosure and to connect it to the *Ge-Schick* (destining).

If it is true, as I believe, that, in his doctrine of the "epochality" of Being, Heidegger reflects and brings to full and rigorous clarity the tendency pervading a large part of contemporary philosophy, then the next question, and the first moment of the ontologie de l'actualité, is, Why should this "declination" of philosophy toward "sociology" (in the rather vague sense that I have given this expression) have become especially visible in our own time? Habermas, in his *Philosophical Discourse of Modernity*, remarks that this "declination" has characterized philosophy since Hegel, but he does not offer a specific and theoretically strong explanation of the fact.[3] Again it is probably Heidegger who suggests the most adequate interpretation of this tendency, an interpretation that can be summarily expressed by pointing to the double meaning of the genitive in the expression "ontology *of* actuality," or "philosophy *of* (post) modernity." The ontology of present existence that we are looking for is a theory that both *speaks about* this existence and *belongs to* this existence. Since, for Heidegger, there is no way of grasping Being beyond its actual event, its specific historical disclosure, a theory that speaks about present existence can only be a theory also based upon this same existence, having no other source of legitimization. In other words, the content of the ontologie de l'actualité that we are attempting to construct must be developed by taking into account the meaning of the needs expressed by this very attempt, by the "declination" of philosophy toward sociology.

The reasons for this "declination," as far as I can see, are not primarily internal to philosophy or rooted in its disciplinary development. A phenomenon, for example, like the debate about *Geistes-* and *Natur-wissenschaften* at the end of the nineteenth century, which in my hypothesis is deeply related to the "sociologization" of philosophy, was not merely an epistemological debate. What was at stake then were the effects of the scientific-technological rationalization of society, against which philosophers and "humanists" wanted to defend an area of freedom that could not be scientifically described or predicted in general laws and might consequently be subtracted from the domination of technology. The atmosphere from which the *Geisteswissenschaften* debate drew its significance was the same that gave birth to the artistic avant-garde at the beginning of the twentieth century (think of Ernst Bloch's *Geist der Utopie*, 1918), the dialectical theology of Barth and Gogarten, the Kierkegaard renaissance, and the increasing "sociologism" of philosophy.

This sociologism and the need for an ontologie de l'actualité that it expresses were reactions to the danger seen in the "total organization" of individual and social life, which starts to be a macroscopic phenomenon in the years around the First World War. The threat involved in this process is not simply the apocalyptic danger of the destruction of individual freedom, feelings, and so forth in the universal functionalization of industrial mass production. It is also the progressive loss of the capacity to grasp a "continuous" meaning

3. Jürgen Habermas, *The Philosophical Discourse of Modernity*, trans. Frederick Lawrence (Cambridge: MIT Press, 1987), pp. 392–93, n. 4.

amid the differing roles that each individual plays in his different social functions. The objectification of individuality within the universal functionalization of the social machinery presupposes and intensifies the fragmentation of the meanings actually experienced by each individual. Heidegger's appeal for the recollection of the sense of Being, which in his hypothesis has been forgotten by metaphysics and by its last and most complete realization, technology, is deeply motivated by these considerations.

To recognize this motivation—of which Heidegger himself became clearly aware only later (which is, ultimately, the meaning of his *Kehre*)—already constitutes a first step in the direction of the overcoming of metaphysics, which, in Heidegger's view, is required in order to satisfy the "needs" expressed by the "declination" of philosophy toward sociology. From the point of view of the disclosure of Being, which belongs to (and to which belongs) the human condition of late modernity, the need for a first principle, which characterized the metaphysics of the past, becomes the need for a "notion" of Being that allows us to recompose a unitary meaning of experience in the epoch of fragmentation, of the specialization of scientific languages and technological capacities, of the isolation of spheres of interest, of the plurality of social roles, and so forth—the epoch, in sum, described by Max Weber in his sociology of modernity.

POSTMODERNITY AS THE ONTOLOGICAL SENSE OF TECHNOLOGY

The philosophical interest in technology—to summarize briefly— is an aspect of a general tendency of contemporary philosophy to conceive of itself as "sociology," or theory of modernity. This trend expresses the effort of the culture at the beginning of this century to counteract the dangerous effects of the progressive rationalization of society, which brought with it specialization, fragmentation, loss of "unitary" meaning, and, consequently, loss of freedom. Heidegger's call for the recollection of the meaning of Being expresses this attitude and is not understandable simply as a development of the epistemological implications of the debate between *Natur-* and *Geistes-wissenschaften*.

Insofar as the specialization of languages and spheres of interest, as well as the fragmentation of existence, characterize modernity (a plausible supposition, supported by many writers such as Max Weber), then the recollection of Being, which is the basis for the reconstruction of a unitary meaning of existence, is also an overcoming of modernity as such. Or, in Heidegger's terms, which I adopt here, it is an overcoming of metaphysics, because metaphysics is at the basis of the scientific-technological framework that supports the modern fragmentation of existence. From the very beginning (Parmenides, Plato), the effort of metaphysics to reach the *arché*, the first principle, was moved by a will to dominate the totality of the world. Through the development of philosophy and science in the history of the West, this purpose has become increasingly concrete and effective: the rational order of the world, which for centuries had been merely "supposed" by metaphysics, has now, at least in principle, been made real in modern technology. Yet this realization brings metaphysics to its end—not only because its "project" is now realized but also, and above all, because the actual rationalization of the world through science and technology shows the real meaning of metaphysics: the will to power, violence, and the destruction of freedom.

The situation in which the need for the overcoming of modernity and of metaphysics becomes urgent is also that in which the violent essence of metaphysics is totally revealed. But this means that we shall not be able to overcome modernity by means of metaphysics, that is, through any foundational thought. The ambiguity of the very term "postmodern" is related to this fact. Postmodern is both a normative ideal and a descriptive, or at least interpretive, notion.

As I have said, within the framework of the epochal essence of Being (which of course implies the multiplicity of epochs, of *Ge-Schick*, which cannot be taken as a metaphysical description of the totality of Being), thought has no other legitimization than the actual disclosure of Being. This determines the peculiar character, and the specific risk, of Heidegger's thought (which has been so overemphasized by Adorno in the chapter "The Ontological Need" in his *Negative Dialectics*): once we have recognized that metaphysical foundationalism is the very basis of the modern fragmentation and violence from which we want to escape, we can no longer attempt to overcome modernity by means of a "critical" thought, which must have at its disposal another *arché*, another first principle. If a postmodern or postmetaphysical thought and form of existence are to be possible, this possibility has to appear within the very framework of late modernity itself. All that thought can do is simply correspond to an appeal that makes itself audible within the same situation that has to be modified or overcome. This assertion involves the risk that philosophy might become mere apology for the existent, as Adorno has said. But it is a risk that cannot be avoided, unless one wants to give up the whole purpose of overcoming metaphysics. In light of this, we can now see that the *ontologie de l'actualité* must take the form of an attempt to discern, *within* the disclosure of Being that characaterizes modernity, the traits of a new disclosure that, among other things, might overcome the specialization and fragmentation of modernity through the reconstruction of a unitary sense of existence.

Where, then, are we to look in order to grasp the present disclosure of Being? This question is just another formulation—again, a postmetaphysical one—of the traditional problem: Where are we to look for the most fundamental truth? Saint Augustine's *in te redi*, Descartes's *cogito*, Kant's conditions of the possibility of experience are some of the most decisive answers to this question. All of them were metaphysical answers insofar as they assumed that truth and Being are stable or eternal. But if one follows Heidegger in his radical thesis of the epochal essence of Being, it becomes impossible to indicate the privileged places where the truth of Being makes itself visible. In a passage of "The Origin of the Work of Art," Heidegger has risked a "list" of the "places" where the truth of Being discloses itself.[4] There appears to be a striking analogy between this list and Kant's philosophy of the faculties, as well as with Hegel's forms of the Absolute and of objective spirit: art, religion, philosophy, moral choice, and political foundations may easily be recognized under the expressions Heidegger uses in that text. That he did not develop these premises, however, and actually limited himself to the dialogue of philosophy with art and language as "ins Werk setzen der Wahrheit" may be interpreted, in my view, as an implicit acknowledgment that the epochal essence of Being ultimately excludes the possibility of establishing once and for all a "table of categories," a systematic definition of the "points" or types of events in which, from time to time, the disclo-

4. "One essential way in which truth establishes itself in the beings it has opened up is truth setting itself into work. Another way in which truth occurs is the act that founds a political state. Still another way in which truth comes to shine forth is the nearness of that which is not simply a being, but the being that is most of all. Still another way in which truth grounds itself is the essential sacrifice. Still another way in which truth becomes is the thinker's questioning" (Martin Heidegger, "The Origin of the Work of Art," in *Poetry, Language, Thought*, trans. A. Hofstadter [New York: Harper & Row, 1971], pp. 61–62).

sure of Being takes place. These places also differ in relation to the difference of the disclosures.

Assuming, then, that we have to look in a direction specific for our epoch, the only indication we possess is that derived from the philosophical texts in which the need for overcoming modernity toward a postmodern, postmetaphysical thought actually expresses itself, namely, the works of Heidegger and those of the philosophers who have oriented philosophy toward sociology in the first half of our century. Again, Heidegger seems to summarize—and radicalize—these indications when he concentrates his reflection on modernity and postmodernity or postmetaphysics on the theme of the *Ge-Stell* (enframing). This word—which, like *Ge-Schick*, takes advantage of the German prefix *Ge* to indicate "a set, complex of, or ensemble"—means, for Heidegger, the totality of the modern rationalization of the world on the basis of science and technology. *Ge-Stell* is the totality of the *Stellen,* of the im-positions; it is metaphysics in its final phase as will to power, as the effective rationalization of reality through reduction of the latter to a system of causal connections dominated by man.

It is in the *Ge-Stell,* Heidegger believes, that the disclosure of Being characteristic of modernity summarizes itself; and it is also in the *Ge-Stell* that an overcoming of metaphysics is announced. In a crucial page of *Identity and Difference* (which, I admit, is almost an *hapax legomenon* in Heidegger's works) he says explicitly that in the *Ge-Stell,* which brings the most extreme danger for man's humanity, we can also see "a first shining of the *Ereignis* (event)," of the (new) event of Being on which depends the possibility of an overcoming of metaphysics. In the ensuing lines, Heidegger explains that this "shining" occurs by virtue of the fact that in the *Ge-Stell* (or in its late modern form) "man and world lose for the first time the characters which had been given them by metaphysics," those of subject and object.[5]

The indication given by Heidegger in this passage is very sketchy and poor. (In a conversation, Hans Georg Gadamer confirmed to me that when Heidegger first delivered the lecture that contains this passage, he himself was perfectly aware of the novelty that this indication represented in relation to his previous works. Gadamer also said he believed that what Heidegger regretted as the real "failure" of his thought was not—as we all know, unfortunately—his political error

5. Heidegger, *Identität und Differenz* (Pfullingen: Neske, 1957), pp. 25–27. In English: *Identity and Difference,* trans. Joan Stambaugh (New York: Harper & Row, 1969).

of 1933, but rather the insufficient elaboration of the relation between his effort to overcome metaphysics and the interpretation of modern science and technology.) What constrained Heidegger to leave this all-important indication in such a sketchy form is probably the same as that which, in the end, prevented Adorno from understanding the potentially positive, emancipatory implications of the mass culture made possible by the media. Both Heidegger and Adorno remained stuck in a vision of technology dominated by the image of the motor and of mechanical energy. Envisioned in this manner, technology could only give rise to a society dominated by a central power that sends its orders to a purely passive periphery, be those orders mechanical impulses, commercial advertisement, or political propaganda.

But, in fact, if we ask more precisely how the *Ge-Stell* might represent an opportunity to overcome metaphysics—to dissolve the subject-object relation that marks human existence in the metaphysical age—the sole possible answer seems to reside in a change of our very idea of technology. It is not the technology of the engine—with its rigid, unidirectional movement from the center to the periphery—that holds out the chance of dissolving the subject-object relation, but rather the technology of communication: the techniques of collecting, ordering, and distributing information. To put it very simply, the chance of overcoming metaphysics, which Heidegger points to rather obscurely as an aspect of the *Ge-Stell*, really displays itself only when techonology—or at least the leading, hegemonic technology—ceases to be mechanical and becomes the technology of information and communication, which we might also call "electronic technology." This development of the indication given in *Identity and Difference*, although obviously beyond Heidegger's immediate intentions, is nevertheless not arbitrary, since it is suggested by other decisive passages in his works, namely, the two essays included in *Holzwege:* "On the Age of the World Picture" and "What are the Poets For?" Particularly in the first of these essays, Heidegger describes our period of increasing specialization in science and technology as the age of the world picture(s): our effort to dominate the world through calculation, now extended to all fields of nature and life, shows an irresistible tendency to dissolve the objectivity of the objects into abstractions. The picture of the world created by natural science tends to explode into a multiplicity of images precisely because of the ever more refined specialization of each scientific language (which also

involves a multiplication of interests, a plurality of the will to power).
Now if one simply extends this thought, the multiplication of world
images, from the natural sciences to the more general sphere of social
communication—the technologies of the press, radio, TV, and now
all that goes under the name of "télématique" (in the word of a recent
French author)—it becomes a little clearer what Heidegger might say
about the capacity of the *Ge-Stell* to accomplish an overcoming of
metaphysics through the dissolution of the subject-object relation
that has dominated the modern disclosure. It would seem that the
Ge-Stell can show the "first shining" of a new epoch of Being precisely
insofar as it brings with it a dissolution of the realistic structure of
experience, a weakening of the "principle of reality," as I will call it.
And, in all probability, it is only the transition of technology from its
modern, mechanical phase to that of electronics, information, and
"mediatization" that makes possible the new dawn of postmodernity,
as well as the philosophical comprehension of what Heidegger meant
in that passage in *Identity and Difference.*

To see this, however, the technology of information must be properly conceived, as it is not, for example, in the simplistic and apocalyptic view of Adorno. It is, of course, true, on the one hand, that
the mass media tend to create ever greater homogeneity and uniformity in the social mind. But no less clear is the opposite phenomenon:
precisely in those societies where the power of the media is greatest,
minorities and subcultures of every kind tend to acquire increased
visibility, even if only to gratify the market's need for novelty and
"differences." The intensification and generalization of the media system also tend to increase the number of "interpretive agencies"; and,
through a paradoxical logic of self-demythologization, these agencies
show themselves more and more clearly *as interpretive.* Think of the
attention the media give to stories concerning the ownership or control of the media, the buying and selling of this or that publishing
company, and so forth. The same is true for the "star system": today
a media or movie star is created and sold along with the explicit story
of that creation, including all sorts of information about marketing,
promotion, monetary success, image, and even the cosmetics that
create the star's "beauty." (In these conditions, there is at least less
danger that anyone will commit suicide when a star dies, as happened, for example, with Rudolph Valentino.)

The open exhibition of the multiplicity of interpretive agents that,
taken together, fabricate our image of the world has the effect of

reducing the "sense of reality" as such: the world is less and less something "out there" and more and more a sort of residue, a crystallization of the "conflict of interpretations" (to use Paul Ricoeur's expression). This "mediatization" of our minds, moreover, acquires its full epochal significance when seen in connection with the dissolution of objectivity in science (referred to by Heidegger in the essay mentioned above), as well as with the dissolution of the objectivity of history in the methodological self-consciousness of contemporary historiography. The effects upon the media society of the emergence of differences and subcultures can also be seen in the historical consciousness of our time—a result of the introduction of other cultural universes consequent to decolonization and the end of eurocentrism. While "mediatization" has given a voice to a multiplicity of minorities and subcultures—and in so doing has revealed the interpretive character of our image of the world—historiography has become aware of the "rhetorical" essence of our ways of reconstructing past events. Not only are there many ways of recounting history, but "history" as such, the pretended objective substratum beneath these historiographical schemes, is itself the product of a presupposed scheme. It is, in fact, the dominating classes that tend to impose their own view of historical development as the real history (as Benjamin and, before him, Nietzsche have shown).

In order to show if and how all these aspects of the weakening of the principle of reality are connected to the culmination of metaphysics in the *Ge-Stell*, one would have to give a more detailed analysis of the connections between the scientific-technological dissolution of objectivity reported by Heidegger and what I am calling the "mediatization" of society, as well as the historiographical dissolution of objectivity. But I hope that, even in the summary terms I have used here, the thesis is at least clear and plausible.

The Weakening of Reality and the Overcoming of Metaphysics

To conclude, in what sense can we expect that this ontology of the weakening of the sense of reality (of Being itself) satisfies the "needs" that, according to my hypothesis, were expressed by the "declination" of philosophy into sociology and, especially, by the Heideggerian appeal for the recollection of Being and the overcoming of metaphysical forgetfulness? Although Heidegger is very reluctant to speak in moral or axiological terms (indeed, the very term "authenticity" disappears from his writings after *Being and Time,* and even there it has a generally nonmoral meaning), we may assume that when one speaks of the overcoming of metaphysics (be it only in the form of a *Verwindung,* of a distorted acceptance of it),[6] one has in mind a process of emancipation, a sort of way out of what the Hegelian-Marxist tradition has called "alienation." This is also true of Heidegger, although he doesn't want to preach in moral or political terms because this would still involve an appeal to human initiative and the consequent danger of remaining within the scheme of the will to power, subjectivism, foundationalism, and so forth. But without pursuing further these questions of interpretation, I come back to the central issue: how far does the weakening of the principle of reality—which I see as taking place in the transition to postmodernity—constitute a response to the needs expressed by the appeal for the overcoming of metaphysics? The answer to this question, I might add, would also help solve still another unresolved problem: What is the meaning of Heidegger's philosophy for the present time? Or, What do we do with Heidegger?

In very general terms, I would suggest that a "weak ontology," or, better, an ontology of the weakening of Being, offers philosophical reasons for preferring a democratic, tolerant, liberal society to an authoritarian and totalitarian one. Certainly it has at least the negative consequence of stripping the ever-recurrent authoritarian temptation of all theoretical appeal and strength. And this remains important: think of the popularity enjoyed by a thinker like Carl Schmitt in recent European political thought, even on the liberal side. I don't believe that Pragmatist and Neopragmatist arguments are strong enough to support a choice for democracy, nonviolence, and tolerance. This observation, I should add, might not bother a thinker like Richard Rorty, who would never expect philosophy to supply reasons for preferring democracy and tolerance; on the contrary, he seems to expect that our preference for a liberal society will supply reasons or impulses for choosing a nonmetaphysical philosophy or, ultimately, for taking leave of philosophy altogether. In this connection, I would just suggest that the sort of self-psychoanalysis of philosophy that Rorty proposed in *Philosophy and the Mirror of Nature,*

6. See Gianni Vattimo, *The End of Modernity,* trans. Jon Snyder (Baltimore: Johns Hopkins Press, 1989), final chapter.

which was meant to dissolve the metaphysical heritage of our philosophy, is not yet and may never be concluded. We still need an
ontology, be it only in order to show that ontology is destined to be
dissolved. Indeed, by way of a provisional conclusion, and speaking
from a standpoint that takes its inspiration from Heidegger, I would
suggest the following: we need to recollect the sense of Being—and
to recognize that this sense is the dissolution of the strong structures
of "reality" into a world of interpretations—in order to be able to live
with this dissolution without neurosis and without the temptation to
"return" to stronger—both more reassuring and threatening—views
of reality. Again, the popularity of various forms of fundamentalism
within late modern Western societies shows that this task is not superfluous. On the ethical level, I suggest that a "weak ontology" has
consequences analogous to the ethics of Schopenhauer (which was
also a model for Horkheimer in his late essays and, less explicitly,
for Adorno).

Of course, the image of the *Ge-Schick* of Being as oriented toward
the weakening of the cogency of reality, subjectivity, objectivity, and
so forth is itself based upon an interpretation. Hermeneutics, to call it
by its name, is not the true, objective description of the hermeneutical
essence of Being; it is already and inevitably a hermeneutical response
to a *Schickung*, to a sending, which it listens to and interprets. *Desto
schlechter*, or, as Nietzsche would say, so what?

Suggestions for Further Reading

Adorno, Theodor W. *Negative Dialectics*, trans. E. B. Ashton. New York: Seabury
    Press, 1973.
Bloch, Ernst. *Geist der Utopie: Gesamtausgabe*. Vol. 3. Frankfurt am Main: Suhr-
    kamp, 1964.
Foucault, Michel. "What Is the Enlightenment, What Is the Revolution?" in *The
    Foucault Reader*, trans. Paul Rabinow. New York: Pantheon Books, 1984.
Habermas, Jürgen. *The Theory of Communicative Action*, trans. Thomas McCarthy.
    Boston: Beacon Press, 1984.
Heidegger, Martin. *Discourse on Thinking*, trans. John M. Anderson & E. Hans
    Freund. New York: Harper & Row, 1966.
———. *On the Way to Language*, trans. Peter D. Hertz and Joan Stambaugh. New
    York: Harper & Row, 1971.
———. "The Origin of the Work of Art." In *Basic Writings*, ed. David Farrell Krell.
    New York: Harper & Row, 1977.
Lyotard, Jean-François. *The Postmodern Condition: A Report on Knowledge*, trans.

Geoff Bennington and Brian Massumi. Minneapolis: University of Minnesota Press, 1984.

Rorty, Richard. *Philosophy and the Mirror of Nature.* Princeton: Princeton University Press, 1979.

10 Liberal Democracy and the Problem of Technology

WILLIAM A. GALSTON

I

Some time ago I was reading Nahum Glatzer's biography of Franz Rosensweig, the great German Jewish thinker who spent most of the last decade of his life in a state of near-total paralysis. The account of his desperate struggle to communicate by means of a specially arranged typewriter keyboard was immensely moving. That very evening I saw on television an IBM advertisement featuring devices that enable quadruplegics to produce documents quickly and easily by directing a beam of light at a computer screen. I thought to myself, "This is technology at its best, helping us to overcome obstacles to the full expression of our natural powers. How many more marvels of learning Rosensweig could have left us if only these devices had been available to him."

A few days later I reached the end of the book, and I was struck by the manner in which, at the final crisis, Rosensweig's doctors and attendants permitted him to die quietly and with dignity. Turning on the television again that evening, I sat through a long story about the all-out medical intervention frequently used in contemporary American hospitals to extend by a few days or weeks the lives of elderly, obviously terminally ill patients. The tubes, the monitors, the elabo-

229

rate machinery of resuscitation—all this seemed grotesque and appalling, made even worse by the evident pain, weariness, and indifference to continued life of many of its intended beneficiaries.

Thus juxtaposed, these two experiences seemed to reveal the two faces of modern technology: on the one hand, the supplement to nature, making good its deficiencies in a manner guided by its innermost possibilities; on the other hand, a breach of natural limits so striking as to recall the classic definition of the fanatic—the person who redoubles his efforts as he forgets his end. The same Janus-faced phenomena are found wherever one looks: the very technology that permits otherwise infertile couples to experience the incomparable joy of bearing a child genetically theirs also makes possible surrogate motherhood, prenatal sex selection, and other practices that recall, all too forcefully, Aldous Huxley's dystopian vision.

This simple intuitive distinction between the completion and the violation of nature raises, of course, an obvious question: does modern thought contain within itself the resources to support this distinction? Or rather, in appealing to it do we fall into a contradiction between the perspective of human experience and the outlook of modern scientific theory, on which, inter alia, modern technology evidently rests?

To elucidate this question, I can find no better point of departure than some remarks offered by Leo Strauss nearly three decades ago in the course of distinguishing between classical and modern political science:

> The Aristotelian distinction between theoretical and practical sciences implies that human action has principles of its own which are known independently of theoretical science (physics and metaphysics) and therefore that the practical sciences do not depend on the theoretical sciences or are not derivative from them. The principles of action are the natural ends toward which man is by nature inclined and of which he has by nature some awareness. This awareness is the necessary condition for his seeking and finding appropriate means for his ends, or for his becoming practically wise or prudent.[1]

So far, this thesis is perfectly straightforward. But there is, Strauss immediately notes, a complication:

1. Leo Strauss, "An Epilogue," in *Liberalism Ancient and Modern* (New York: Basic Books, 1968), p. 205.

Prudence is always endangered by false doctrines about the whole of which man is a part; prudence is therefore always in need of defense against such opinions, and that defense is necessarily theoretical. The theory defending prudence is, however, misunderstood if it is taken to be the basis of prudence.[2]

The application of this account to classical political science is relatively nonproblematic. Aristotle confronted the challenge of ancient materialism, which lent considerable support to political conventionalism at the expense of natural justice. In Aristotle's view, this philosophical materialism was not only practically pernicious but theoretically flawed. He was thus able to advance his physics and metaphysics as both philosophically superior to its predecessors and as supportive of the pretheoretical experience of human ends.

It hardly needs saying that since the onset of modernity, matters have grown more complex. A new natural science based on mathematical materialism emerged victorious in its struggle with the Aristotelian scientific tradition. It is ridiculous to deny that this victory was in some substantial measure merited. Whatever its other motives and effects may have been, and whatever its ultimate limits turn out to be, the new natural science represented—and continues to represent—a marked gain in human knowledge. This gain did not, however, come without cost. In the seventeenth century it was already understood that, applied to our species, mathematical materialism could well entail substantial revisions in the understanding of human psychology, action, and morality—changes that might require the abandonment of much of what had previously been regarded as normative bedrock.

There were four main lines of response to this new situation. The first, exemplified by Thomas Hobbes, among others, was simply to accept these implications—that is, to accept mathematical materialism as the monistically adequate account of the whole and build a new science of human practice on that basis. This effort ran into some well-advertised difficulties, vividly exemplified by Hobbes's forced resort to a distinctively human linguistic capacity whose relation to generative material motion is at best obscure, midway on the path from matter to politics.[3] Ever since, a range of human phenomena—

2. Ibid., p. 206.
3. For an extraordinary discussion of this difficulty, see Strauss, "On the Basis of Hobbes's Political Philosophy," in *What Is Political Philosophy?* (New York: Free Press, 1959), pp. 174–82.

chief among them consciousness itself—have proved stubbornly resistant to reinterpretation in the language of mathematical materialism. That modern science can provide important insights into the
material substrate of consciousness is clear enough, but the experience of consciousness seems to go beyond the account of its causes
and preconditions.

These and related difficulties have rendered scientific monism im-
plausible, although it by no means lacks advocates even today. A
number of alternatives have surfaced during the past two centuries.
At one extreme lies what might be called the intentionalistic reinterpretation of the scientific enterprise. On this view, modern natural
science is not—was never—oriented toward truth but rather to specific practices and ends. It was intended to expand human power over
nature, or (what is not necessarily the same thing) to facilitate the
use of nature to relieve the historic miseries and infirmities of the
human condition, or even to increase the plausibility of atheism by
defining out of existence, in advance, the possibility of miracles.

If these alternatives are considered simply as historical propositions, there is something to be said in favor of each of them. But as
a total account of modern science, they are defective and one-sided
for two reasons. First, as I have indicated, science has, in fact, aug-
mented our knowledge, not just our power. Over the past four centuries, our understanding of the cosmos, organic as well as inorganic,
has steadily increased. It is one thing to affirm that this understanding is less than comprehensive, or even that it will always remain less
than comprehensive; it is quite another to deny its very existence.
Second, if, as Aristotle suggests, the desire to know is natural to our
species, it is most unlikely that this impulse simply disappeared half
a millennium ago. I would suggest, rather, that natural science is one
of its contemporary expressions. As Hans Jonas puts it, "There is still
'pure theory' as dedication to the discovery of truth and as devotion
to Being, the content of truth: of that dedication science is the modern form."[4]

The third response to the rise of modern science was the Romantic:
the effort to reinterpret scientific procedures and results in less
mechanistic and mechanical, more aesthetic and teleological terms—
that is, to restore something like the Aristotelian unity of theoretical
and practical science on terms more hospitable to human experience

4. Hans Jonas, "The Practical Uses of Theory," in *Philosophy and Technology,* ed. Carl Mitcham and Robert Mackey (New York: Free Press, 1972), p. 346.

than the orthodox interpretation of modern science would permit. The first wave of this effort, foreshadowed by Rousseau and grounded in Kant's *Critique of Judgment*, came to fruition in the scientific writings of Goethe and Hegel. It is generally, and I believe properly, regarded as a failure. More precisely, it either repeated—and reified—the existing state of scientific knowledge in somewhat more obscure terminology, or it speculatively deviated from that knowledge, with by and large unfortunate results.

Some contemporary versions of this strategy seem distinctly more promising. As set forth by such thinkers as Roger Masters and Leon Kass, the core thesis is that modern biology can be partially detached from physics, and that, understood as quasi-autonomous, the new biology can provide support for the distinctiveness of the human species and the meaningfulness of natural ends as they present themselves in ordinary experience. Kass asks, "Should not the remarkable powers of self-healing, present in all living things, make us suspect that dumb nature in fact inclines purposively toward wholeness and is not simply neutral between health and disease?"[5] Masters states flatly that "contemporary natural science is not automatically hostile to a doctrine of natural right based on the assumption that living beings differ in decisive ways from inanimate bodies."[6] At this juncture, we do not know whether this attempt to rejoin natural science with Aristotle (which would, of course, set aside the anti-Aristotelian animus driving modern science from its inception) will succeed. If it does, a centuries-old difficulty will be overcome, and contemporary practical philosophy will be able to proceed far more straightforwardly and securely.

But as long as this reinterpretive strategy is in doubt, we will be forced to consider the fourth major response to the rise of modern science—namely, the dualism of human consciousness and material extension initially proposed by Descartes and brought to its fullest exposition in Kant's first and second critiques. In this view, it does not really matter how mechanistic, antiteleological, or reductionist natural science turns out to be, for the simple reason that it does not—cannot—characterize what is distinctively human. Our inward awareness of reason, of free and responsible action, of the essential

5. Leon R. Kass, *Toward a More Natural Science* (New York: Free Press, 1985), p. 8.

6. Roger D. Masters, "Evolutionary Biology and Natural Right," in *The Crisis of Liberal Democracy: A Straussian Perspective,* ed. Kenneth L. Deutsch and Walter Soffer (Albany: SUNY Press, 1987), p. 60.

unity of our selfhood through the myriad of sensations and emotions—all this remains untouched by even the most perfect account
of the external world in the language of efficient causation.

The dualistic response to science encounters two familiar difficulties. First, our understanding of what is distinctively human cannot
but be affected by our understanding of the nonhuman. It is no accident that Kant's mechanistic understanding of the natural world co-
exists with a freedom-centered account of humanity. More generally,
heteronomy would cease to be such a threat, and autonomy an absolute requirement, if the laws of otherness were not seen as so inhospitable to our distinctiveness. Second, dualism inevitably raises the
question of how heterogeneity can coexist within a single, coherent,
noncontradictory universe. For Descartes this question takes the form
of the "mind-body problem"; for Kant it emerges as the mysterious
nexus of free human action and material motion in a necessitarian
world.

At a critical juncture, Kant's dualism rejoins Aristotle's more unified
view. For Kant, as for Aristotle, our understanding of our humanity
takes, and must take, our everyday prephilosophic experiences at
its point of departure. These experiences, moreover, enjoy a kind of
independence from theoretical philosophy, in the sense that their
force in no way depends on their deduction from physical or metaphysical premises. They can, however, be endangered by false theory,
or even by true theory that oversteps the bounds of its validity. Properly understood, theoretical inquiry not only engenders truth within
its proper domain but also defends practical philosophy from theoretical error and excess. In a famous passage Kant declares that "innocence is indeed a glorious thing, but on the other hand, it is very sad
that it cannot well maintain itself, being easily led astray. For this
reason, even wisdom—which consists more in acting than in knowing—needs science, not to learn from it but to secure admission and
permanence to its precepts."[7]

This has been the hastiest of sketches, but it suffices, I believe, to

---

7. Immanuel Kant, *Foundations of the Metaphysics of Morals*, trans. Lewis White Beck (Indianapolis: Bobbs-Merrill, 1959), p. 21 [Akademie 405]. I do not mean to imply that Kant and
Aristotle had an identical view on this matter. For Aristotle, everyday practical consciousness
is the beginning of wisdom, but dialectical inquiry can end by revising the content as well
as the form of this awareness. Kant comes much closer to arguing that the dialectic alters
the structure of our moral awareness without affecting its conduct-guiding content.

shed some light on the question that evoked it. In principle at least, modern thought does contain some resources for grounding a distinction between the proper and improper use of technology. If some version of neo-Aristotelian accommodationism is, in fact, validated, then the direct (if untutored) experience of natural human ends will suffice to set in motion a dialectical inquiry that culminates in practical direction for the employment of means. And if some version of Kantian dualism is preferred, then an account of human freedom can help draw the line between technology supportive of our dignity and technology that erodes dignity by treating us as means alone.[8]

II

However this may be in theory, one may wonder whether in practice modern societies have succeeded in keeping technology within appropriate bounds. There is a widespread belief that they have not, that technology has run amok, that it has become an end in itself rather than a means, even that it has become the inescapable basic structure within which all of modern life transpires.

A particularly revealing version of this thesis was advanced decades ago by Martin Heidegger. The "inner truth and greatness" of national socialism, he asserted, concerned "the encounter between global technology and modern man." This is so because "from a metaphysical point of view, Russia and America are the same, the same dreary technological frenzy, the same unrestricted organization of the average man."[9] In Heidegger's view, as Catherine Zuckert observes, the primary cause of national socialism's ultimate failure was its absorp-

8. There is, of course, a considerable distance between these theoretical perspectives and specific practical controversies. It might be argued, in fact, that the distance is so great as to render these theories all but useless. But this view is at once too demanding and too pessimistic: too demanding because theory typically gives us a vocabulary and grammar for discussing practical disputes, not straightforward recipes for their resolution; too pessimistic because theory can strongly incline us to favor one solution rather than another. For example, a neo-Aristotelian conception of species-normal functioning would enable us to distinguish gene splicing designed to remove congenital defects (e.g., Down's syndrome) and promote normal fetal development from genetic engineering undertaken to alter the character of the human species. I am grateful to Robert Kraynack for the questions that provoked these reflections.

9. Martin Heidegger, *Introduction to Metaphysics*, trans. Ralph Manheim (New Haven: Yale University Press, 1959), p. 37.

tion into the reigning ideological-technological competition among the great powers.[10]

Heidegger's argument represents the inadvertent self-undermining of any effort to characterize modern politics from a "metaphysical point of view." The reason is straightforward. While there are no doubt similarities among fascism, communism, and liberal democracy, these resemblances are at most secondary. Today, as in classical antiquity, the ways in which communities choose to organize themselves and the goals they choose to pursue are fundamental. It makes a decisive difference whether the technology of war is used to liberate or to enslave, whether the technology of medicine is used to relieve human misery or to impose it, whether the wonders of modern pharmacology are used to ameliorate mental illness or to induce it. If the twentieth century has proved anything, it is that politics is still the frame of human existence, and that liberal democracy is politically superior to its major competitors.

That this is so is evident in the sphere of technology. As a political matter, the limits to what might be termed the totalization of technology have proved far more effective in liberal societies than in other highly technologized orders. Setting aside the extreme examples of Nazi death camps and Soviet psychiatric hospitals, the fact is that the contemporary environmental movement began in the liberal democratic West, where it has found far more effective political expression than anywhere else. By contrast, the recent opening of Eastern Europe has highlighted the environmental catastrophe wrought by the heedlessness of its command economies. One might also note that, whatever one's views of the merits of nuclear power, rising public concern has been translated into effective restraints on its development to a far greater extent in the United States than anywhere else. Chernobyl was significantly more lethal than Three Mile Island, but its domestic political consequences for the regulation of technology are as yet far smaller.

Let me broaden this point. The core error of Heidegger's thesis is its inability to distinguish among political orders and to give appropriate weight to the differences. This error was in turn rooted in Heidegger's studied refusal to look at human things from a human point of view. Consider the following statement, uttered in 1949: "Agriculture is now a mechanized food industry. As for its essence, it is the same

10. Catherine H. Zuckert, "Martin Heidegger: His Philosophy and His Politics," *Political Theory* 18 (February 1990), 71.

thing as the manufacture of corpses in the gas chambers and the death camps."[11] To this we must reply: No, it is *not* the same thing. Using tractors and fertilizer to grow food is not the same as using railroads and poison gas to exterminate human beings. That both are "technological" tells us only that what Heidegger regards as the "essence" of the matter is in truth quite secondary.

Political inquiry is, or ought to be, a kind of humanism. By this I mean that its central concern should be the effect of politics, and of politically relevant phenomena, on the well-being and flourishing of human beings. By humanism I also mean the determination to take seriously, at least as our point of departure, the concrete conceptions of (and judgments about) well-being and its opposites offered by real human beings in the course of their daily lives. It is from this standpoint that liberal democracy, for all its well-advertised flaws, has emerged as decisively superior to its competitors. Specifically, it has offered freedom from state oppression, physical security and material decency for most (though not all) citizens, a closer approach to distributive justice made possible by abundance, and unprecedented liberation of inquiry and of the human mind from political coercion. It is the credibility of these promises, coupled with the bitter experience of their opposites, that has led the peoples of Central Europe to reject state socialism and to embrace liberal democracy as their best hope for the future.

III

To argue, as I have, for the continuing primacy of the political is not to say that there are no similarities among contemporary technological societies, or (what amounts to much the same thing) that there are no systematic differences between antiquity and modernity, taken as a whole. To explore these differences, we may usefully focus on the differences between the Greek conception of *technē* and the modern experience of technology. As a summary judgment, it seems evident that although Aristotle's depiction of *technē* as an ensemble of tools and know-how for the fulfillment of human purposes remains valid up to a point, it fails to capture the totality of contemporary technological phenomena. The following familiar points seem to me to characterize the core of the difference.

11. Quoted in Zuckert, "Martin Heidegger," p. 71.

To begin with, Aristotle presents *technē* as substantially detached from *epistēmē*. In modernity, by contrast, technology stands in the closest possible relation to natural science, and most of the major technological changes—from electrical energy to atomic energy to computers—have been made possible by leaps in basic science.

Second, Aristotle proposes a sharp distinction between the activity of philosophy, which is contemplative, and the activity of *technē*, which is transformative. In modernity, as has been evident since Bacon, "natural philosophy" rests on systematic interaction with the objects of investigation. To be sure, the point of this mode of inquiry is theoretical in something like Aristotle's sense—to allow nature to reveal itself as it is. This is why Francis Bacon, along with the mainstream tradition of natural science down to the present day, insists that obedience to nature precedes the capacity to command nature. Still, the relationship between *technē* and *epistēmē* has been transformed, so much so that one may even speak of a circle: scientific advance makes possible technological change, which in turn makes possible further scientific advance. For many scientific purposes today, the quality of instrumentation is critical (consider supercomputers, electron microscopy, the proposed supercollider), instrumentation that rests squarely on the immediately preceding generation of basic science.

Third, and perhaps most important for present purposes, Aristotle's depiction of *technē* implies (if it does not quite say) that human ends are fully controlling and that *technē* can be understood as pure means. The contemporary experience of technology calls this into question. To begin with, modern technology turns out to have intrinsic qualities that shape human activity and political possibility. Whether we examine the modern corporation as a system of production, or the modern automobile as an object of action and consumption, the quality of the means considered in itself cannot be disregarded. For example, driving a car requires a specific ensemble of skills and awareness and tends to evoke a unique set of emotions. And, as many social theorists have observed, it alters the experience of travel by determining what we can do, and notice, as we travel.

Related to this is the fact that the adoption of specific technologies not infrequently produces unexpected consequences, above and beyond the ends explicitly in view. Who anticipated in 1918 that the postwar automobile would become a vehicle not just of physical loco-motion but also of sexual liberation?

More broadly, the commitment to a specific technology usually turns out to entail a commitment to its implicit logic—that is, to the system that will enable it to function at ever-increasing levels of effi-cacy. Thus, for example, the increasing social centrality of the automo-bile led to Robert Moses' wholesale reconfiguring of New York City, stunningly depicted by Robert Caro, and to the creation of the inter-state highway system, which spurred economic growth along the corridors it established while strangling hundreds of small towns that found themselves off the beaten track.

Finally, of course, technology takes on the aspect of autonomous development, outside the control of human agency and political deci-sion. This appearance of autonomy, I suggest, is created by the logic of two kinds of competition—military and economic. As has been evident since classical antiquity, advances in military technology can confer real advantages on their possessors, and other political com-munities may well feel compelled to match them. (The alternatives are to hope that the intentions of the more advanced power are be-nign, or to entrust the security of one's community to an equally advanced ally.)

More recently it has become evident that economic growth is heavily dependent on constant technological innovation, which is engen-dered most reliably by competitive markets. But economic competition is not a cab one can stop wherever one wishes. To seek refuge from—or to ignore—competitive pressures is, inevitably, to fall behind. Thus the sad but well-merited fate of the American auto-mobile and steel industries. Thus also the much-remarked fact that the sheltered economies of Eastern Europe and the former Soviet Union have become technological museums. This does not amount to an argument from historical necessity. A community that is willing to exchange an increment of better living standards for what it re-gards as a more than compensating increment of stability may reason-ably choose to restrict, or opt out of, the processes of economic competition and economic dynamism. My point is only that very few peoples in modern times have knowingly made such a choice, and that by selecting instead the objective of rapid economic advancement they have committed themselves to a social process that takes many subsidiary decisions out of their hands—whence the subjective expe-

rience of technological autonomy and concomitant social powerlessness.

IV

On the basis of the foregoing, I may now restate the guiding questions of this inquiry. Does technology, so understood, promote or
inhibit the flourishing of liberal democracy? Conversely, does liberal
democracy contain within itself the intellectual, social, and institutional resources to deal appropriately with modern technology?

To address these questions it is necesary to begin, in a few broad
brushstrokes, by offering some conceptual clarifications and clearing
away some long-standing misapprehensions. To define terms, let me
characterize the liberal democratic polity as a community possessing
to a high degree the following features: popular-constitutional government, a diverse society with a wide range of individual opportunities and choices, a predominantly market economy, and a substantial,
strongly protected sphere of privacy and individual rights.

One long-standing misconception about liberal democracy is that
it can be understood in terms of processes and institutions without
reference to ends. This misconception has been given new life during
the 1980s by liberal theorists such as John Rawls, Ronald Dworkin,
and Bruce Ackerman, who have argued that the core of liberalism is
the widest possible neutrality concerning social ends, particularly the
good life and what gives value to life. While not without surface
plausibility, the neutrality thesis cannot be sustained.[12] Whether we
consider the history of liberal thought or the actual functioning of
contemporary liberal institutions, we find instead a rough consensus
on a handful of goals and goods that give liberal public life its point.

In my book *Liberal Purposes*, I have ventured to enumerate this
consensus in the following terms. Liberals believe that life itself is
good and that the taking or premature cessation of life is bad; that it
is good to be born with normal, as opposed to defective, basic human
capacities and to be able to develop them normally; that it is better
to fulfill than to stifle legitimate individual interests and purposes;
that freedom, properly understood, is an indispensable element of

12. For a sketch of the reasons why, see William A. Galston, "Defending Liberalism,"
American Political Science Review 76 (1982), 621–29, and "Pluralism and Social Unity," *Ethics*
99 (July 1989), 711–26.

each individual's good; that without a reasonably robust capacity for rationality, individuals will be hard pressed to function in liberal economic, social, or political life; that opportunities for a rich network of human connection and association are vital; and that needless pain and suffering are to be minimized.[13] No doubt other formulations of this consensus are possible, but I strongly suspect that they will end by covering much the same ground.

A second long-standing misconception about liberal democracies is that they can function well simply on the basis of self-interested individual behavior unguided by anything more than institutionally structured incentives. This "invisible hand" conception of liberalism is not true, and was never true, even of competitive markets; it is certainly not valid for liberal democracy considered as a whole. The liberalisms of Locke, Kant, and Mill, while diverging in many respects, agree on this: liberalism is unimaginable without the general acceptance of certain core virtues. The insight of theory is amply confirmed by practice: liberal democratic economies, societies, and polities depend heavily on the widespread practice of the liberal virtues.

Some of the virtues needed to sustain the liberal state are requisites of every political community: the willingness to fight if necessary on behalf of one's country, the settled if not completely determinative disposition to obey the law, and loyalty—the developed capacity to understand, to accept, and to act on the core principles of one's community. Some of the individual traits are specific to liberal social life— for example, independence, tolerance, and respect for individual excellence and accomplishments. Others help sustain the liberal economy: for instance, the work ethic, which combines a sense of obligation to support personal independence whenever possible through gainful effort with the determination to do one's job thoroughly and well. Still others are entailed by key features of liberal democratic politics. For citizens, the disposition to respect the rights of others,

13. Galston, *Liberal Purposes* (Cambridge: Cambridge University Press, 1991), chap. 8. For an earlier (and, I now think, less adequate) version, see my *Justice and the Human Good* (Chicago: University of Chicago Press, 1980), chap. 3.

It may be thought, as G. A. Cohen has suggested to me, that these elements of the good are too general to be specific to liberalism. I am certainly prepared to allow that, as a theoretical matter, there is considerable common ground between liberalism and democratic socialism. If we ever see an actual example of a democratic socialist regime, we can then determine the extent to which, as a practical matter, this conception of the human good is consistent with it. Eastern European, Soviet, and Chinese socialism clearly have not promoted the liberal good as I understand it.

the capacity to evaluate the talents, character, and performance of public officials, and to moderate public desires in the face of public limits are essential. For leaders, the patience to work within social diversity and the ability to narrow whatever gap may exist between wise policy and popular consent are fundamental. And the developed capacity to engage in public discourse and to test public policies against our deeper public convictions is highly desirable for all members of the liberal community, whatever political station they may occupy.[14]

V

To insist, as I have, that liberal democracy is structured by certain ends and virtues is not to say that these ends are invariably pursued or that these virtues are reliably practiced. Indeed, the virtues I have enumerated are the ones liberal democracy needs, not necessarily the ones it has. In light of these facts, we may rephrase our questions yet again. What is the impact of technology on the attainment of liberal goals and the development of liberal virtues? To what extent have liberal polities been able to bring technology under the control of their central concerns and purposes, and, conversely, to what extent has technology tended to undermine them?

Let me begin at the incontrovertible beginning. The development of technology within market economies has made possible unprecedented abundance, which in turn has enabled liberal polities to preserve and extend life, to relieve suffering, to expand the range of personal choices, to allow many (though by no means all) of its citizens to develop their capacities through education, training, and occupational diversity, and in general to enjoy a greater measure of freedom from the burdens and constraints that in all previous ages have afflicted our species.

Our technology-driven abundance has also made possible a substantial amelioration of distributive injustice. Here Strauss's account is particularly relevant. During classical antiquity, education and training were necessarily confined to an elite (the "gentlemen") because scarcity precluded their wider distribution. From one important

14. For the arguments underlying this brief summary, see my "Liberal Virtues," *American Political Science Review* 82 (December 1988), 1277–90.

standpoint, this restriction was blatantly unjust. It violated the basic principle that equals should be treated equally, for those who received a gentlemanly education were by and large no more talented, no more innately equipped to use it well, than were those who did not receive it. The resulting aristocracy was conventional, not natural, and could be sustained, if at all, only on the dubious grounds of social utility. Scarcity forced a choice between injustice and a "state of universal drabness," and injustice at least gave "some grace and some distinction" to even those at the bottom of class-divided societies.[15] But with increasing abundance, Strauss continues, "it became increasingly possible to see and to admit the element of hypocrisy which had entered into the traditional notion of aristocracy." The new guiding principle was instead equality of opportunity to develop and display whatever gifts one might have. This meant the abolition not of natural inequality but only of its unjust conventional simulacrum: "natural inequality has its rightful place in the use, nonuse, or abuse of opportunity in the race as distinguished from at the start." Thus, he concludes, "it became possible to abolish many injustices or at least many things which had become injustices."[16]

It is, of course, possible to argue that the benefits of abundance have come at too high a cost; this has been a familiar theme of social criticism at least since Rousseau. There is little doubt that the increased freedom of citizens in wealthy societies has, somewhat paradoxically, been accompanied by increased dependence (real as well as psychological) on the technological systems and objects of consumption. And for a substantial portion of the past two centuries, a plausible case could be made that even if the fruits of technological abundance were widely distributed, the human costs of engendering it were excessive—a charge for which the dehumanizing effects of the division of labor typically constituted Exhibit A.

But we should hesitate to render a verdict without considering a fuller range of pertinent evidence. To begin with, it is easy to romanticize the character of preindustrial labor, a temptation that few critics of technological capitalism have been able to resist. More important, there is growing evidence that the postindustrial division of labor is rather more hospitable to human development and dignity than was the industrial system it is rapidly replacing. Employers report (and

---

15. Strauss, "Liberal Education and Responsibility," in *Liberalism Ancient and Modern*, pp. 11–12.

16. Ibid., p. 21.

the market confirms) a dramatically increased demand for well-educated and trained workers who are capable of responding flexibly and responsibly to a wide range of nonroutinized challenges. The fears, so widely bruited only a few years ago, that the postindustrial work force was in the process of becoming "deskilled" are almost certainly misplaced.

There is mounting evidence, moreover, that the demands of economic efficiency and workplace practical intelligence are growing steadily more compatible. A widely discussed case is the NUMMI plant in Fremont, California, a highly successful Japanese-run venture that replaced failed General Motors management at an aging, inefficient, and strife-torn automobile plant. When asked to explain his secret formula, the Japanese president replied, It's simple. GM treated its workers like machinery; we treat them like human beings. We can't maintain our competitiveness without constant improvements in production processes, and we've learned that workers know more about that than anyone else. It is necessary to consult them in the manner that respects their dignity and evokes their cooperation by showing them that they, too, have a stake in the implementation of the suggestions they are making.

Yes, it may be countered, but this is to take much too narrow a view of the matter. Is it not the case that technological abundance is leading inexorably to the degradation of culture and of human beings? A review by Thomas Molnar offers a representative sample of this thesis. He writes that modern man

> is so conditioned [to fight against the breach of his political rights] that he hardly notices the greater threat of the uniformity, homogeneity, and quantification of his existence; as he fights for the possession of a wide array of machines in his private life, he puts shackles on his imagination and freedom; as he insists on increased consumer goods and services, he lets himself be tied to networks of electronic apparatuses which give him instructions, fill his head with useless information, cut away the edges of his personality, curtail his inner self. One may legitimately wonder if what Tocqueville dreaded: the tutelary state gently removing "the citizens' burden to think and to live," may not be more justifiably feared on the part of our technological paradise?[17]

17. Thomas Molnar, "Is Technology Ideological? The Other Face of Politics," *Political Science Reviewer*, 1989, 295–96.

The very familiarity of this litany may induce us to nod in assent, and surely elements of it are recognizable in our ordinary experience. Yet Molnar's concluding reference should make us wonder about the underlying thesis. In Tocqueville's view, after all, it was not technology but rather democratic equality—the abolition of aristocratic distinction and privilege—that unleashed the desire for ever-expanding material abundance and imparted a curious homogeneity (or, as Tocqueville put it, "monotony") to democratic society. Twentieth-century technology has surely not retarded this process. On the other hand, Tocqueville was already able to say a century and a half ago that "variety is disappearing from the human race; the same ways of behaving, thinking, and feeling are found in every corner of the world."[18]

I would suggest that the complaints summarized by Molnar are not so much misguided as misplaced. It should never have been imagined that liberal democracy is so capacious as to include within itself every human possibility, or that the transition from monarchy or aristocracy to liberal democracy would be wholly without cost. It is natural to lament the submergence of aristocratic distinction in mass society; it is natural to contrast aristocratic independence of mind and action with the relative uniformity of liberal democratic denizens, but (as Tocqueville recognized) it could not have been otherwise. The fundamental question is whether, taken as a whole, liberal democracy is superior to the forms of political and social organization it superseded. If, as I believe, the answer to that question is clearly affirmative, then we should stop wringing our hands about what was, after all, an explicit clause in the original bargain. This does *not* mean that we should cease thinking about ways of reintroducing an awareness of true aristocratic excellence at the margins of democratic society; it does mean that we should abandon the illusion that this awareness can ever become a dominant force in such a society.

I have discussed the impact of one great sphere of technological competition—the economic—on liberal democracy. What about the other—military competition? Here I can be brief. The technologizing of warfare has clearly rendered all-out international conflict far more destructive than ever before. At the same time, by raising the stakes so high, it may well have lowered the odds of such conflict very significantly. We simply do not know what the vector sum of these

18. *Democracy in America*, vol. 2, pt. 3, chap. 17. This entire chapter deserves a more careful consideration than is possible within the compass of this essay.

opposed tendencies will be. We do know that the technological balance of terror coincided in Europe with an unprecedented period of tranquillity and eventuated in an opening (fragile, to be sure) toward democracy. But we cannot be sure of either the causal connection or the ultimate outcome.

Of two other matters we can be somewhat certain. First, the prolonged competition in military technology has not produced what many feared—a thorough militarization of liberal democracy in which the armed forces and intelligence services would end by subverting civil liberties and constitutional government. Serious breaches have surely occurred, but in the main, over time, liberal democratic institutions have proved equal to the challenge. Second, the transfer of military technology to domestic society, with all its negative effects on personal security and domestic tranquillity, is thoroughly relative to political culture and under the control of public institutions. Great Britain and Canada have murder rates far lower than that of the United States, in large measure because they have chosen to crack down on the personal possession and use of high-technology weapons of destruction. The critical variable is politics, not technology itself.

The effects of technology on the core liberal democratic principles of equality and freedom are complex but not, on the whole, negative. As already indicated, technology makes possible a closer approach to liberal social equality—meritocratic equality of opportunity—than was conceivable in economies of scarcity. It also facilitates the overcoming of obstacles to the expression of natural gifts—that is, it helps diminish the consequences of certain kinds of infirmities. We may well be on the threshold of a new and more troubling relation, in which technology can directly intervene to interrupt and correct a range of genetically programmed misfortunes. Here we return to what might be called the Rosensweig conundrum. It is one thing to use genetic engineering to nip Down's syndrome in the bud, that is, to eliminate natural defects in a manner that safeguards the life of the fetus. It is quite another to use this capacity to increase the incidence of certain natural traits at the expense of others, or even to produce new traits never before seen in the human species.[19]

19. That this developing capacity is profoundly troubling goes almost without saying. But it is a misunderstanding to suppose that an increase in our power to manipulate the genetic constitution of our species somehow represents a denial of human distinctiveness. On the contrary, it could be said that humans are the only species with the power to affect con-

Since at least the early nineteenth century, fears have been wide-spread that technology would contribute decisively to liberal democratic inequality by tilting the social and political balance in favor of those with advanced technical knowledge, paving the way for an orgy of paternalistic social engineering. On balance, I would suggest, these fears have not been borne out, both because liberal communities have exerted political control over technology when they felt it necessary and because there has been a fairly steady shift in liberal democracies against the appropriateness of public paternalism. Indeed, at the level of both moral theory and institutional practice, liberal democracy has erected far more effective bulwarks against such manipulations than have any of its modern competitors.

This is not to say that contemporary technology poses no threat to political liberty. The rapid development of computer systems, for example, has made possible an unprecedented pooling of information and has given centralized government agencies broad new capacities to monitor the activities of individuals. In these trends a disruption of long-established checks and balances and a breach of traditional divisions between the public and the private may well be seen. At the same time, the development of personal computers has tended in the opposite direction: the decentralization of the capacity to obtain, to interpret, and to disseminate information. I, for one, find it extraordinarily difficult to draw a compelling conclusion from these countervailing trends, all the more because computerized centralization can itself serve important liberal democratic purposes—for example, the more reliable detection of wealthy tax evaders who were previously able to conceal or leave unreported substantial dividend and interest income. Similarly, advanced telephone technology, which has done so much to invade privacy, can also be employed to protect it: witness the development of devices revealing the telephone numbers of incoming calls, the threat of which constitutes a dramatic deterrent to cranks and sexual harassers.

VI

Thus far I have offered a relatively sanguine view of the relation between modern technology and a healthy liberal democracy. But

---

sciously the direction of change in its own genetic makeup. Technological advance does nothing to undermine this distinctive capacity; the question is rather how we use it.

there is a darker side as well. Let me conclude by discussing what I take to be four serious problems created, or exacerbated, by techological advances.

First, as I argued earlier, liberal democracy rests on a collective commitment to maximize the preservation of human life and minimize the pain and suffering associated with life. The strength of these ends as individual motivations requires little public reinforcement. But if they are to be effectively collectivized, we must be able to take an interest in the lives and woes of others, and to do that, we must be able to extend to them some measure of sympathy and compassion. On balance, I would suggest, modern technology reduces our capacity to do this by distancing us from others and by transforming death and suffering into electronic stories that we no longer experience as fully real. Video games and cartoons are the standard examples, but murder reports on the local evening news are not much better. An extension of this occurs in military combat. Thomas Molnar pertinently cites the high-altitude destruction of Coventry, Dresden, and Hiroshima, when the pilot-technician dropped the bomb on an invisible and (to him) no longer human target, as examples of the way technology can distance the emotional and moral self from the consequences of his acts.[20]

Second, technology does not create, but it surely exacerbates, the vice of acquisition without limit. This form of behavior was already well known to classical antiquity. In Book 1 of the *Politics*, for example, Aristotle discusses this phenomenon, which he sees as endemic to commercial occupations, and links it to the boundless desire for self-preservation and sensual gratification.[21] Modern liberal democracy, of course, publicly liberates these desires to a greater extent than ever before and gives pride of place to commercial activities based on technologically driven economic growth as the way of satisfying them. But at the same time, liberal societies cannot remain healthy without at least a moderate individual and collective capacity for delaying gratification. For if this is lacking, many individuals will find it difficult to save adequately by curbing current consumption and to invest sufficiently in long-term education and training, and the polity as a whole will find it difficult to adopt policies that balance future interests appropriately with present concerns. The easy availability of endless new technological products has added significantly to these

20. Molnar, "Is Technology Ideological?" p. 301.
21. *Politics* 1.9.1257b30-1258a13.

difficulties in recent decades, and it is far from clear whether it will prove possible to mobilize effective countervailing forces.

A third problem is the impact of technology on the pace of social change. In a famous passage, Aristotle distinguishes between progress in the arts and sciences and progress in the law: "Change in an art is not like change in law; for law has no strength with respect to obedience apart from habit; and this is not created except over a period of time. Hence the easy alteration of existing laws in favor of new and different ones weakens the power of law itself."[22]

Aristotle's point may, I believe, be broadened beyond the law. Healthy human development and existence require a measure of stability, that is, a framework within which habits may be formed and expectations satisfied. Excessively rapid change contributes to a sense of disorientation, a loss of efficacy, a feeling of disconnection and deracination. While liberal democracy historically has embraced optimism about change to a greater degree than most other forms of political organization, it is hardly exempt from the psychological generalizations just sketched.

I think a strong case can be made that, in part under the impact of intensified international economic competition driven by relentless advances in technological products and processes, the pace of social change in liberal democracies may now exceed the capacity of many individuals to maintain their inner balance. With regard specifically to the United States, the point is this: to an extent not previously seen, Americans must now choose between enhanced economic competitiveness coupled with pronounced social dislocation and slower long-term economic growth that promises a somewhat more manageable pace of social change.

The consequences of technology for social change are but a subset of a larger problem, toward which I can do little more than gesture. Liberal democracy is the sturdiest product of the Enlightenment faith in the congruence between the progress of the human mind and the progress of society. Not surprisingly, the community of scientists has furnished an enduring ideal for liberal democratic public life: a model of inquiry among equals, oriented toward truth and human betterment rather than toward power, inwardly protected against the inevitability of individual prejudice and bias, insulated from nefarious external interference, carried out in conditions of fullest possible pub-

22. Ibid., 2.8.1269a20-24.

licity, and ultimately determined only by the force of the better argument.

We need not dwell on the extent to which actual scientific practices have, in fact, diverged from this noble vision. The critical fact for our purposes is that the very nobility of the scientific ideal has lent continuing support to the thesis that the public fruits of the scientific enterprise would naturally nurture both the public institutions and the private desires of liberal democracies. The difficulty we encounter in discussing public limits to technological dissemination is traceable to the enduring power of this assumption. Challenging it might in the end engender a richer and more cogent defense of liberal democracy at its best. But in the short run, such a challenge would inevitably call into question much of what we cherish most in modern civilization.

The final problem concerns the impact of communications technology on liberal democratic politics.[23] Let me preface this discussion with a general observation. Americans are accustomed to regarding technological innovation as expanding our choices, as in many ways it does. We are less aware of the respects in which innovation does not add to, but rather displaces or drives out, previous possibilities. The dissemination of the telephone, for example, has all but destroyed civilized correspondence.

This observation applies with particular force to contemporary politics. The ubiquity of television has dramatically diminished the power of other forms of political communication. Recent studies have shown that, as a percentage of adult population, newspaper readership has sharply declined over the past three decades. A 1988 survey revealed that 84 percent of all Americans now regard television as their single most significant source of political information.

Many elected public officials and representatives of the media are deeply concerned about this development; rightly so, in my judgment, for it means nothing less than the replacement of words with images, of the ability to concentrate with the desire for quick hits, of the awareness of logical and causal relations with the craving for jump cuts. On balance, television has made the serious discussion of ideas, the serious conduct of campaigns, the serious consideration of alternative public policies far more difficult than ever before. Various correctives are now surfacing: restrictions on thirty-second advertise-

23. Here I speak simply as an American; I do not know enough about the mass media in, e.g., Italy or Britain to judge whether my observations are of more general application.

ments, legally required extended segments of free television time during general elections, the acceptance of more meaningful debate
formats as a condition of publicly financed campaigns. Ross Perot's
thirty-minute issues commercials suggest that a more informative use
of television can hold the public's attention. If widely adopted, these
strategies may foster much-needed improvements. But they will stand
in constant tension with the underlying tendencies of a predominantly visual medium.

In 1990 Philip Roth offered an account of an extended discussion
with the Czech dissident novelist Ivan Klima. Roth concluded his side
of the conversation with a sober warning:

> As Czechoslovakia becomes a free, democratic consumer society,
> you writers are going to find yourselves bedeviled by a number of
> new adversaries from which, strangely enough, repressive, sterile
> totalitarianism protected you. Particularly unsettling will be the one
> adversary that is the pervasive, all-powerful archenemy of litera-
> ture, literacy, and language. . . . I am speaking about that trivializer
> of everything, commercial television. . . . At long last you and your
> writer colleagues have broken out of the intellectual prison of Com-
> munist totalitarianism. Welcome to the World of Total Enter-
> tainment.[24]

Roth is making an extremely important if not wholly unfamiliar
point. State repression is the enemy of democratic discourse, but so
is the endless desire to be entertained. Public dialogue has satisfac-
tions of its own, but it takes effort; it is work. A public that demands
constant entertainment is by that very fact debarred from meaningful
participation in the serious business of democratic self-government.
But this we already knew from the Roman historians: bread and cir-
cuses was the motto not of the republic but of the empire.

24. Philip Roth, "A Conversation in Prague," *New York Review of Books*, April 12, 1990,
p. 22.

Suggestions for Further Reading

Borgmann, Albert. *Technology and the Character of Contemporary Life.* Chicago: Uni-
versity of Chicago Press, 1984.
Durbin, Paul T., ed. *Research in Philosophy and Technology 7.* Greenwich, Conn.:
JAI Press, 1984.

Winner, Langdon. *Autonomous Technology: Technics-Out-of-Control as a Theme in Political Thought*. Cambridge: MIT Press, 1977.
———. *The Whale and the Reactor: A Search for Limits in an Age of High Technology*. Chicago: University of Chicago Press, 1986.

11 Technology and the Problem of Liberal Democracy

JERRY WEINBERGER

I

The title of this essay suggests not so much that technology is a problem as that in our technological age something is wrong with liberal democracy. Immediately two objections come to mind. First, we have just witnessed the astonishing collapse of communism as a political system and as an object of moral aspiration. With the exception of a few third-world stragglers, all countries now seem to be groping toward some form of liberal democratic capitalism. It is hard to deny that technology was important if not decisive in assuring this great revolution. Although the powers of religion, nationalism, and democratic spirit were also important causes, they pale before the model of liberal individualism and prosperity, grounded in technological prowess and disseminated by the technological shrinking of the world. Courage and sacrifice were certainly displayed, and one cannot dismiss the steadfast American leadership of the free world. Yet these more noble and political virtues could not have been sucessful without the extraordinary achievements of science and technology: the nuclear balance of terror created a world of such stability that the alternatives to liberalism were defeated by their failure to deliver on the modern project for conquering nature. It is hard to

253

explain the rebirth of freedom for the peoples in the East without considering their (and their former Soviet masters') discovery that the shopping was so much better on the Western, liberal side. So how could technology highlight a problem with liberal democracy?

Second, it would seem that liberalism is the only serious frame of reference for managing the problem of technology as we have come to know it. In its usual formulation, this problem is said to be the danger that with our newfound technological prowess we will, despite good intentions, deform our very humanity. Technology enhances our powers of manipulation, while modern science inclines us to think of human nature as subject to the blind laws governing the behavior of matter in motion. Under the spell of these two contradictory forces we risk the possibility of self-inflicted dehumanization and brutalization, all in the name of human comfort, freedom from pain, and variety of choice. If this is the danger of technology, one doubts that we could confront it apart from the moral categories of modern liberalism: the dignity of the individual, the sanctity of human rights, economic liberty and free markets, cultural and political pluralism, and representative democracy. If technology threatens our humanity, it does not becloud our understanding of just what that humanity is—about that there is an apparently universal, and liberal, consensus.[1] Of course technology is a problem, but it is not fundamental since it has not obscured the liberal moral stars by which we take our bearings.

There is much force in these objections, but they ignore a curious fact. We now hear it said that both technology and liberalism are exhausted at the very moment of their triumph—that we are entering a postmodern era beyond the moral, scientific, and political categories of modern rationalism. Much of this talk is superficial and trendy, the grumbling of intellectuals who cannot stand to see the victory of the hated bourgeoisie. Still, one cannot deny that it reflects the strange mood of uneasiness that has come upon the victorious West, the anxiety that grips cultural and political elites as well as more common folk at a time when one might expect jubilation or smugness. And at its bottom is a very serious argument rooted in Martin Heidegger's account of technology. Despite the bad odor of his politics, Heidegger's argument about technology has greatly determined the course of modern thought in general, and on our shores it has

1. So universal that history has been declared to be at its end. See Francis Fukuyama, "The End of History?" *The National Interest* (Summer 1989), 3–18.

influenced recent interpretations of liberal politics. Thus we cannot ignore Heidegger's claim that as technology stamps our age both technology and triumphant liberalism fall into crisis.

In Heidegger's view, modern technology is not the totality of machines and techniques, which we must tolerate, destroy, or master as a merely neutral set of means lest we forsake our humanity. Technology is not a collection of material powers we need to do something about. Rather, it is the thinking that forms modern life. Moreover, this thinking begins with obfuscation and ends with a shattering opening of the eyes. The technological frame of mind proceeds to its own undoing, a catastrophe for reason but a liberation for human experience.

To speak very generally (and perforce to distort quite a lot), what Heidegger means is that the modern world is fashioned and lived according to reason, whose virtue is exemplified by objective natural science. But modern science is in essence technological and thus counts as "real" only the predictable, the calculable, the manipulable. Thus in defining our epoch, scientific reason distorts the phenomena of human life. In a technological world reverence and faith, loyalty and community, art, and the full possibilities of astonishment and estrangment are lost and covered over. The technological age is characterized by nihilism, by the view that what does not fit the ontology of technological reason simply *is* not. For nihilism, what is not explained by science is subjective value, not objectively real but merely relative to some arbitrary, historical situation. And most important, in the nihilistic age everything—whether objective, subjective, or divine—is treated as the object of manipulation and use. Scientism, relativism, and historicism, and the ruthless exploitation and use of the world and everything in it, are all aspects of nihilism. Both communism and liberal democracy—with their relentless need to conquer nature, subdue the gods, and rationalize human life—are products of nihilism.

But nihilism eventually destroys itself. It ultimately reveals and undermines its own hidden source, which is the dogmatic conception of Being at the heart of Western rationalism. This happens as follows: according to the technological world view, the phenomena of life as it is experienced are not objective. But it is only a short step from such a view to the realization that objective reason is itself a way of life and thus without rational grounds. The ontology of objective reason (but not the positive truth of science) is exposed as a faith or fate, and when this light dawns, reason's distorting grip on life and Being is broken. We then see that Being is not a knowable and manipulable object. Being reveals itself as the finite, mysterious, unpredictable source of all that happens. We can then see—without dogma and without the drive merely to use and manipulate—how experience is in thrall to powers beyond our control. Being is now grasped as the awesome source of open historical destiny. The drive for technological mastery thus compels us toward the strange power of fate; the urge to calculate prepares the way for genuine surprise; the defeat of theology reveals new gods; and the corrosive power of relativistic individualism frees the call of new communities. This is what Heidegger means by saying that technology, whose essence is "by no means anything technological," is a danger that "saves."[2]

Now as we shall see, Heidegger's teaching has been absorbed by two schools of political thought in America: one could be described as democratic-liberal and the other as progressive-liberal. Despite their differences, they share the Heideggerian view that the problem of technology has nothing to do with the unintended consequences of particular inventions, however vexing such consequences can be. They accept Heidegger's claims that the problem of technology is nothing technological, that it rather concerns a way of thinking, and that the real promise of technology is the self-destruction of that way of thinking. For both schools, the demise of technological rationalism has important implications for our liberal order: either it calls for a democratic transformation, or it forces us to rethink what liberalism requires of us and, for the sake of liberal progress, to admit that until now we have been mistaken.

II

We must first consider what these schools have to say, beginning with the more radical and democratic critique. While recognizing that the term is an amalgam that obscures subtle and important differences, I refer to the several authors of this critique as the "participatory democrats." They are liberals to the limited extent that they reject communism, fascism, and nationalism and accept essentially liberal

2. Martin Heidegger, "The Question concerning Technology," in *The Question concerning Technology and Other Essays*, trans. William Lovitt (New York: Harper & Row, 1977), pp. 28–35.

conceptions of individual autonomy and human rights. But they argue that the technological demise of rationalism makes it possible to rediscover genuine politics, the real stuff of life that is obliterated by both liberalism and communism.[3] For the democrats, the political realm must be reclaimed if we are to achieve wholeness in our lives. The problem is not with material technology as such but with life formed by technological rationality or "instrumental reason." The enemy is not the gadget but manipulative rationality embodied in the practices and concepts of liberal politics: abstract individualism, representation, the distinction between the state and civil society, constitutional checks and balances, disciplined two-party government, and, perhaps most important, bureaucracy and bigness.[4]

According to the democrats, these liberal forms are devices that sever politics from the rest of life. Liberal politics is thus truncated politics in which the noble activity of practical deliberation is distorted and transformed into an instrument of oppression and degradation. Politics—and thus life—is stifled by the liberal distinction between the state and civil society, especially by the welfare-state interventions in civil society that do occur in the liberal order. For these interventions take place at such an abstract level that they serve only to mask manipulation and control.[5] Civil society is not a community of whole persons but is rather formed from such reifications as "consumer," "producer," "voter," "householder," "manager," and so on. Liberal politics dismembers human beings into theoretical abstractions, which are then reassembled into manipulable constructions that impoverish the soul. Such politics—which is justified as providing for the "pursuit of happiness" and unlimited material pros-

3. Both liberalism and communism are products of technological rationalism prior to its nihilistic self-destruction.

4. The "participatory democrats" are, of course, a broadly mixed group, with some, such as Sheldon Wolin, more influenced by the Frankfurt School and others, such as Benjamin Barber and the anarchist Reiner Schürmann, more directly moved by Heidegger by way of Hannah Arendt. Moreover, if one includes among the democrats the communitarian interpreters of Rawls, such as Michael Sandel, the "democratic" spectrum itself runs from left to right, even though it would be correct to think of it as in general on the left. For the purposes of this essay I ignore these differences, however important they are, for two reasons: first, in America the Frankfurt critique of technology has been itself strongly colored by Heidegger (Wolin, for instance, is as influenced by Arendt as by the Frankfurters); and second, my intention here is to raise questions that arise from the common ground staked out by the democrats regardless of their significant differences. For the latter purpose, the amalgam I describe does not distort the relevant facts.

5. See Wolin, "Democracy and the Welfare State: The Political and Theoretical Connections between *Staatsräson* and *Wohlfahrtsstaatsräson*," *Political Theory* 15 (Nov. 1987), 467-500.

perity—is technological through and through. In theory liberalism protects the autonomous individual's right to mind his own business. But in practice, liberalism causes the important decisions affecting life to be made by external forces, according to principles of cybernetic rationality that are blind to genuine human ends.

Now the democrats do not deny the liberal distinction between public and private life, but they do claim that this distinction can be healthy only when these realms of life are much more fully integrated than they are in our technological politics. For the sake of justice and moral health, the public life of the citizen must be as concrete, immediate, and extensive as are private concerns. Only then can the fragmentation of individual and community be avoided. The problem with our age is not material technology, which cannot be outstripped, but politics conceived as technology rather than as *praxis* that autonomously constructs the genuine moral community. In the final analysis, the participatory democrats are revolted by the kind of human being supposedly produced by liberal rationalism: a morally denuded creature for whom there is no possibility of meaningful place and location, no possibility of loyalty and reverence, no possibility of genuine participation in common life, and thus no possibility of genuine creativity. In the world of such "individuals," most have ample opportunities for being oppressed and all suffer deformation of the soul.

The hope of the democrats is thus to fatten up the thin, liberal self. But it must not be allowed to grow too fat or be defined too rigidly by some fixed and larger whole. This would be to give up the goal of genuine individual autonomy just because the liberal form of autonomy is a sham. It would be to succumb to the abstract, disintegrative "totalizing" of technological rationality, which is, as we learn from Heidegger, the common bond among communism, fascism, and liberal democracy. On the one hand, life must transpire in whole, robustly political communities, but on the other hand, the individual must not be constrained by the irrationalities associated with premodern life—the oppression of women, the intolerance of civil religion, the inequalities and stupidities of tradition, not to mention the glorification of war and aggression. The proper balance can be struck, but only when the harmful rule of technology has been overcome.

This hope need not be scoffed at as unrealistic. As Heidegger shows, the collapse of utilitarianism is the very destiny of technology itself, not some silly romantic dream. And the collapse of technological rationality will not eliminate the apparatus of material technology; it does not mean a return to the flint and the horse-drawn plow. We can have our cake and eat it, too, in the following sense: when technology frees us from its own distorting frame of mind, we can then take full advantage of its material bounty. We can construct the community as a moral whole and yet be free from the material scarcity and related habits of mind that until now have made communities oppressive.[6]

The last point discloses the real issue between traditional liberals and their democratic critics. The liberal view is simply that the democratic conception of the good life—the participatory fashioning of the common good—is much too partisan. In order to be fully democratic, the good life must not be constrained by inequality. Genuine citizenship, as the democrats understand it, would require near perfect equality since differences in means become impediments to full and free participation. The democrats cannot advance their view of the good life without at the same time advancing a partisan view of justice. Regarding the most fundamental (but not the only) political disagreement, they take the side of the many poor against the few rich.[7] The liberal thus concludes that one cannot regard democratic political life as the highest good without endangering moderation and freedom. The long and bad reputation of direct democracy—that it results either in civil war or imperialism—was well earned by democrats and well learned by liberals.[8] Whatever the good life may be, says the liberal, it must be private, with politics both conceived of and practiced as a subordinate means. The effect of such subordination is social and political inequality because wealth and political skill are among the goods pursued and generated in private, and because in the absence of artificial intervention, natural inequality and luck will have free rein. But such inequality is first of all preferable to tyranny and internecine conflict and second of all subject to moderate reform.

6. There is a difference between the more political democrats who recognize the necessity and inevitability that any political whole must make exclusive demands and claims and the anarchists, who do not. The anarchists would accuse the former of wanting to have their cake and eat it, too, and of being too close to the earlier, more dangerous Heidegger. The more political democrats would accuse the latter of refusing to own up to the full range of possibilities that make up political life. Here I am thinking about the more political democrats, although my considerations would be relevant—with suitable changes in emphasis—to the anarchists as well.

7. Cf. Aristotle, *Politics* 1279a26-1281a11.

8. Even for Aristotle, the most democratic of the three healthy regimes, polity, is based on the courage of an armed many.

Democratic talk about the inherent goodness of political life is just partisan enthusiasm, cloaked in a smoke screen of fancy theory about the moral and ontological superiority of *praxis.*

The democratic response is that in our technological age partisanship is not really such a danger, and that once this fact is recognized, the genuine attractiveness of political life will become evident. Under conditions of material plenty, it is possible for the rich and the poor by themselves to reconcile their differences—in the direction of the poor—without resorting to war and imperial expansion. Moreover, to believe this is not to suffer the delusion that the material scarcities about which partisans disagree can be wholly overcome. On the contrary. Even the democrats understand that, in the absence of disagreements about material shares, it would be hard to see what political life would be about, or how it could matter at all. But the democrats do maintain that political participation can become a virtue that gives transcendent value to deliberation about material shares. For this to be possible, competition must be urgent enough to matter and yet sufficiently mild that it need not be vicious. Under such conditions, competition can be gilded by the virtue of common deliberation.

Thus the democrats hold that virtue requires scarcity that does not press too hard. In other words, democratic virtue requires that something other than virtue should work in favor of the many poor. And here the democrats have a good case against the liberals, for that "something" is what the liberals themselves count on so heavily: the productive power of technology. In this light, the liberals show up as the unreasonable partisans. In a situation of plenty, where genuine democratic citizenship is actually possible, the liberals invoke the bogey of political conflict to preserve inequalities that they profess to condemn, that are in fact perverse, and that stunt the really fulfilling possibilities of human life. The liberal is deluded by technological rationality that sees life as never complete, as forever pushed harshly by the need for technical ingenuity. In fact, technology gives us all the material things we require for democratic life, and it undermines its original frame of mind that sees needs where they do not really exist.

The real issue between democrats and liberals thus concerns the nature of technology, both as regards its material effects and the way in which it discloses the world of human experience. Ultimately, and with Heidegger, the democrats argue that technology overcomes *technological* politics. If liberals could just see what is in front of their

noses—the essence of the technology on which they themselves depend—they would give up their fear of politics and recognize the
deformities of soul and body politic that come with the liberal state.
There is really no need to ask for any favors, however, because technology is a force with a logic and will of its own. The age of mature
technology is perforce a postmodern era. And in this new era we bid
farewell to liberalism and the technological frame of mind.

The more liberal critique of liberalism is posed today by those I
will call the "progressives," whose position is best expressed in the
work of Richard Rorty.[9] They have no quarrel with liberal politics and
suspect that the democrats are romantic at best and authoritarian
at worst. But having learned about technology from Heidegger, the
progressives think liberalism is misinterpreted by fuddy-duddy liberals, those conservative Western hard-liners who think that liberalism requires some demonstrable foundation for its conception of the
human good. Conservative fogies worry that modern nihilism (scientism or existentialism) will shake these foundations, without which
the liberal house will fall. And so they rush to defend Enlightenment
rationalism, or natural right, or Christianity, or even the humanities
(the Canon). But for the progressives, all this worry about foundations and nihilism is the fretting of Chicken Little.

Nihilism might be dangerous—and thus might require the mythology of foundationalized morality—if it resulted in the rise of some
dogmatic cultural and political authority. This once happened with
scientific positivism in the English-speaking world and Heidegger's
fascistic attempt to reinterpret the philosophical language of the West.
But in these earlier cases nihilism was not fully mature, as it is in our
time. For us, nihilism—the gift of technology—makes it impossible to
take such big claims at all seriously. As regards stultifying positivism,
mature thought now shows that there is no rational warrant for taking
science as the only way of life. As technology forces the project of
reason to full, hermeneutical self-consciousness, the scientific paradigm undoes the very authority by which it could do any serious
damage. And as regards Heidegger and his right-wing historicism,
we need not worry about the return of some strutting militarism.
What holds for positivism does likewise for any political implications

9. Richard Rorty, *Contingency, Irony, and Solidarity* (Cambridge: Cambridge University
Press, 1989). Rorty speaks for a broad range of left-leaning liberals, especially academics
influenced by postmodernism and the concern for diversity.

of fundamental ontology. (In this respect, the early Heidegger is just the final gasp of rationalism, and the later Heidegger pulls the young one's sharp, still metaphysical teeth.) The real danger was always the lure of metaphysical absolutes, whether they appeared nakedly as religion or more subtly as idealism, positivism, or existentialist fascism. This danger has passed now that reason's undoing of every absolute has become part of our morning cereal.

The progressives think that if there is a problem with liberals, it is simply that some have not yet come to grips with this end of foundationalism. They have yet to embrace fully their postmodern, cosmopolitan urbanity, which grants a human solidarity that reduces every communal boundary to a flexible aesthetic creation, satisfying to the individual imagination but making no hard demands, conferring no (spurious) depth of soul, and harming no one not included. For liberals to clear their heads, they need only accept the genuine gift of technology as Heidegger has interpreted it. They need only recognize the dawn of the postrationalist, postmetaphysical age. And then the story and the character of liberalism appear in an entirely new light.

In this new light, liberalism appears as a way of life that separates our "final vocabularies" (conceptions of the ends of life) as much as possible from public life, thereby minimizing the possibility of coercing the whole of a community on the basis of some partial and contingent view about ends. This separation between public and private life is not grounded in some natural truth but is simply the political system required by the liberal's abhorrence of cruelty—the vice that most represents minding another's business more than is necessary. Until recently, it was thought that liberalism was unique in being founded on reason alone. It was thought that if liberalism were to engage our energies in the struggle against its enemies, it had to answer to reason. It thus had to be based on a true account of nature in order to be superior to politics based on religion, superstition, or tradition. Liberalism stood or fell, then, with the possibility of epistemology.

But at the technological end of the Enlightenment era, we know that epistemology was a dream and that such a true account of nature does not exist. Although liberalism at first rode on the shoulders of rationalism, we now understand that this was but faith in reason's ability to ground political life in a true or objective account of the world and the human soul. The legacy of the Enlightenment— moving dialectically and by way of Romanticism and the critique of technology—is the self-overcoming of rationalism, the critical disclosure that reason cannot rationally establish its own grounds, that reason is just another way of life, and that any rational theory about the nature of the world and the self is bounded by contingent presuppositions about the world, about the soul, about Being.

We now know that the particular vocabularies used to grasp the world are given to us as a historical legacy that just happens and that is never the result of some deliberate choice or act of rational discovery. In the big picture, ideas do not matter in the sense of being the causes of historical change: the Russian Revolution would have happened as it did if Karl Marx had never lived, and liberal politics would have developed if there had never been a John Locke.[10] Liberalism—the final vocabulary or story that has come down to us—is simply our own and nothing more. If we find that its political outcomes are decent in comparison with historical alternatives, then we should thank our lucky stars but make no attempt to think that our institutional practices are warranted by a demonstrable morality or rooted in the truth about the world. Yet every historical world is also the product of art (history is not "natural," after all). This is the meaning of the discovery that the root of every world is inherited convention. If we can speak of reality, it is creativity, not truth. But since everything is both creativity and legacy, there is no simply free creativity. Given the fact of historicism, all creativity occurs in the tension between the old and the new. The new can never be free from the burden of the old, from the weight of the given, which is the only anchor for creativity that is essentially finite.[11]

The view that everything is legacy and creativity is not a danger for liberal politics because the technological demise of rationalism spawns a new kind of liberal. This is the character Rorty calls the liberal ironist. The ironist knows that there is no one final vocabulary and that all such vocabularies are at once created and in the grip of history. Armed with this knowledge, ironists are playful, unwilling and unable to take themselves seriously in the sense that there should be significant social or political consequences of thought about final vocabularies. Thus ironists will not be inclined to boss others

10. Of course, as products of history, ideas have very important and powerful local effects. Ideas make a given vocabulary concrete and thus have some limited causal power. But this power just isn't very important in the big historical picture, and nothing we can do by way of thinking will change that picture, which is itself not caused by specific acts of thinking. I am indebted to Arthur Melzer for urging me to clarify this matter.

11. Rorty, *Contingency, Irony, and Solidarity,* p. 73.

around. Moreover, they do not have to be fully self-conscious philosophers.[12] It is true that the philosopher and the artist, unlike the average person, live in the greatest possible awareness of the existential
burden of irony. Lacking aesthetic and philosophic intelligence, the
nonphilosophic liberal would tend rather to act out of habit, after the
manner of Aristotle's conception of moral virtue. But even so, these
more common folks would accept the difference between public and
private and would be historicist ironists, although only commonsen-
sibly.

It is true that no society can have an ironic public rhetoric lest it
socialize its youth "in such a way as to make them continually dubious about their own process of socialization."[13] In fact, then, the common folks would accept their liberal principles as if these final
vocabularies were not contingent and altogether historical. For them,
if they were to think about it, their habitual ironic relativism would
be a genuine theoretical problem. These folks, who are not intellectuals or professors and who actually make liberalism work, will not be
postmetaphysical relativists all the way down. The philosopher-artist,
on the contrary, will be, possessing the courage to live with the contingency of "our" most fundamental world view—that of liberalism
itself.[14] And yet despite this difference in liberal types, there is no
politically important difference between the pedestrian many and the
few intellectuals because the nonphilosophers' residual dogmatism is
in our time simply harmless—"ironized," we might say, if not
strictly ironic.[15]

We have every reason to believe that the legacy of our history is
progress toward the liberal habit of privatizing inherited and recreated vocabularies, whether or not those vocabularies are taken with
full or merely habitual irony. In other words, the intellectuals need
tell no noble lies because the lies merely ironized folks will believe in
(despite the open mockery of them by the intellectuals) are harmless,
held as they are by people who abhor cruelty and mind their own
business. And the people need not fear the intellectuals because,
having recognized their fundamental political irrelevance, the intellec-

---

12. Ibid., pp. 73–95. Not all ironists will be liberals as such, but they will nevertheless
incline toward liberalism since they certainly incline toward liberalism's relativism, distinction
between public and private life, and abhorrence of cruelty.

13. Ibid., pp. 87–89.

14. Ibid., pp. 73–78, 87–88.

15. Ibid., pp. 194–95.

tuals will perforce mind their own business and, in any case, just say more authentically what is implied in how everyone else acts.

In our time, "political freedom [sucessful liberalism] has changed our sense of what human inquiry is good for."[16] In the age of mature liberalism, the purpose of human inquiry—philosophy or thought—is not the quixotic (although perhaps at one time useful) project of grounding an absolute liberal morality because our politics no longer requires it. In the age of mature liberalism, human inquiry owns up to what it always has been: creative art that serves politics because thought does not escape the weight of the given, which is always political. In our time, however, it can serve not by grounding various possibilities of political discipline and exclusiveness (even if they are liberal), but by playfully and privately expressing the easy cosmopolitanism of contemporary liberalism.[17] Indeed, once we admit to the subservience of thought to politics (and history) and can see that thought has no foundational responsibility, that it does not matter in the historical scheme of things, thought becomes free to create (without political effect) in its own private sphere. The dependence of thought on history thus opens up the genuine freedom and autonomy of thought, which is essentially liberal because it is essentially private.

In the late technological age, the activity of human inquiry ceases to have any deleterious political consequences. If one could only ask—not compel by reason—the likes of the illiberal Heidegger (or the hard Nietzsche or the young Paul de Man) to privatize his particular final vocabulary, there would be nothing to worry about because it is simply the case that no one will take such vocabularies seriously.[18] The kind of world in which compelling demonstrations of truth might be thought either essential or useful is one where the creative redefining of the self could possibly annihilate the politics of liberal irony. But this possibility is simply no longer in the cards: the privatized self of liberal democracy is not just the present weight of the given; it is now rather the very horizon within which everything can be grasped as historically given.[19] Thus despite the difference between

16. Ibid., p. 68.

17. The rationalists and the foundationalists, and all those serious about morality, are not really dangerous. Rather they are to be pitied for being too crabby and judgmental to be able to join in on the creative fun.

18. Rorty, *Contingency, Irony, and Solidarity*, p. 197.

19. This horizon is simply a contingent gift of history. No thinking brought it to us, nor is it grounded theoretically in the "truth" of historicism. There is no necessary or causal

the few intellectuals and the many liberal ironists, there is nothing that an artist-philosopher might say that could matter politically because, if anyone pays any attention, it will be consonant with what "we liberals" think—that is, it will be consonant with habitual liberal relativism, the distinction between public and private, and the abhorrence of cruelty.[20] The courage of human inquiry is not really a big deal, since it involves no risk from political or moral indignation.

The relation between thought and politics—between the plain folk and the intellectuals—reflects the character of liberalism in the postmodern age. The key is the absence of indignation on the part of the merely ironized, common many. Far from being a danger to liberalism, the demise of rationalist foundationalism enhances the general ironic mood. Epistemology and rationalist absolutism incline even the liberal to the habit of excluding and harming. This may have been useful in past wars against slavery, fascism, or communism, but these evils are long gone. Thus were we still to think liberalism in need of foundations, our society would be more disciplined as a community, and so crueler and less caring and inclusive and easygoing than in fact it has to be or now is. The liberal order now admits without worry the free power of creativity. In our case there is a compatibility between the "truth" about human experience—the fact of its essential or down-to-the-roots contingency—and our moral intuitions and liberal political practices.[21] The demise of foundationalism thus serves

connection between historicism and the substance of liberal moral intuitions. Again, I thank Arthur Melzer for urging me to clarify this matter.

20. It is doubtless this view that makes Rorty so acute in his reading of the various revolutionary postures one finds in the air today. The worst we need fear from the Nietzscheanized left, who, like conservatives, think philosophy is where the political action is, is that they will take over the university English department (not the government) and from there influence the required reading in the high schools. If they succeed, neither the government nor academic freedom will be different than they are today, but we will be made more sensitive to blacks and gays. "De Man and the American Cultural Left," *Essays on Heidegger and Others*, vol. 2 (Cambridge: Cambridge University Press, 1991), pp. 137–39.

21. To attribute such a truth claim to Rorty is not to set him up for the accusation of self-refutation. He is not guilty of self-refutation any more than is any really thoughtful historicist, fundamental ontologist, or deconstructionist. When Rorty or any historicist says that every account of a world is an interpretation, he is saying (more strongly) that what we can know about the world is never enough to satisfy the ambitions of the metaphysician, for whom truths are really the absolute grounds of practical moral rules, or (more weakly) that every account of the world up to now has been perspectival or grounded in some contingent and ungrounded assumption about the world. While such claims are certainly debatable and refutable, they are not self-referentially incoherent. They could be considered so only by begging the important question: what is the whole, how does it appear, and how can we grasp it? For the most part, the "self-refutation" argument is an excuse for not having to read a hard book.

progress toward greater political freedom and equality: toward the softening of political and communal boundaries, toward pluralism, and toward a liberal ethnos whose ethos is enlargement and diversification.[22]

The participatory democrats and the new progressives are parts of the cloth now woven from Heidegger's understanding of technology. Regardless of their differences with Heidegger and disagreements among themselves, they accept Heidegger's basic argument that clarity about the politics of our current situation comes first and foremost from recognizing the saving self-destruction of technological rationalism. This rationalism can remain as a molted husk, as it does in the case of old-fashioned liberals and to the detriment of human freedom. But its day has passed. The intellectual weight of the metaphysical tradition—with its moralizing, totalizing, and technological ontology—is now being removed from our lives. And this feat was accomplished by the metaphysical-technological tradition itself, by its self-defeating drive to disclose a knowable world to a knowing subject. Technology, as Heidegger said, is a danger that saves. Whatever is now to be done (it could be revolutionary nay saying, small democracy, or just becoming more caring and less judgmental), getting a bead on politics and human freedom requires grasping our new location beyond reason. In this situation, liberalism can never be the same, either because its forms, institutions, and moral vision are simply otiose, as the democrats say, or because it is no longer possible to see its forms, institutions, and moral vision as presenting in any way a genuine problem, as the progressives say.

III

It must be admitted that at present neither the democrats nor the progressives have much practical influence on our liberal institutions.

---

22. We must not think that Rorty is smuggling the causal power of ideas back in through a rear door. Regarding the course of liberal ironism, there is no choice for philosophy, and liberal politics does not depend on what ironic philosophers do or discover. The only real question is whether or not liberal intellectuals make fools of themselves by taking a side that does not really exist. At best, what is at stake is whether one avoids being silly—whether one gets to be the giver rather than the receiver of knowing condescension. One can then answer the moral demands of liberalism—the commandment at once to care and to chill out—by mounting the critique of morality. One then really gets to have one's cake and eat it, too: to doubt everything including morality and be in tune with popular morality. The very ease of holding this position, its irresistible charm and popularity, demonstrates

They are, to put it bluntly, confined to the narrow halls of academic life and the hermetic pages of the intellectual press. But they surely could become influential. Public opinion does seem out of sorts and in search of new directions ("it's time for a change"), and it must not be forgotten how closely liberalism is tied to the project of modern technology, which becomes more doubtful as our mood gets ever "greener." So if the democrats and the progressives have grasped the real destiny of technology, we should look to them for guidance. If they are wrong about technology, then perhaps such views are predictable in a technological age, in which case we will have to take their measure as forces to be reckoned with. It will be useful, therefore, to probe the assumptions and arguments about technology posed in these views: do they announce the saving truth of technology, or do they merely reflect confusions fostered by technological utopianism? These are very large and difficult questions that cannot be settled here. But it is possible to clear the air a bit in order to see what issues are really at stake.

The first step is to avoid starting out too theoretically—too quickly absorbed in abstruse discussions of metaphysics, epistemology, and the essence of the Western tradition. Thus we begin by putting the matter in the simplest and most concrete terms: the difference between old-fashioned liberals, on the one hand, and the democrats and progressives, on the other, is the extent to which political life is judged to be really dangerous business. For old-fashioned liberals, the common good is not produced by a deliberate consensus about the just and the good or by inexorable moral progress. Rather, it consists of the principles and institutions (the most important of which are constitutional devices) that make possible contrived and artful compromises.

In such a scheme the common good—self government and political and individual liberty—is removed as far as possible from substantive claims about the good life. These claims or opinions are relegated to private life. Privatizing the choice of ends makes it easier to take a formal or abstract stance toward the distribution of material shares, toward the question of justice. Those who lose in the economic game have the solace of their political liberty and freedom to mind their own business, which they prefer to any specific conception of distributive justice. Those who lose in the political game (often the same)

---

that postrationalist liberalism makes no serious demands. The progressive ironist is a squared circle: a merely playful, antimorality moralist.

have the solace of being free in private life, which bolsters the freedom to try again in politics. And all have their lives softened by the
rising standard of living produced by free economic competition.

Of course the result of such formal liberties—economic inequality—
places concrete limits on the range of goods that can be chosen by
some individuals in their private lives. Thus the liberal common good
is never simply and unproblematically common. It is in some way
artificial—the product of persuasion, if not coercion, rather than of
some agreed upon and demonstrable principle of distributive justice.
To the liberal this fact is not embarrassing. On the contrary, it actually
recommends liberalism, which manages better than any other form
of government the harsh conflicts that inevitably arise in human life.
For the first liberals, this fact of life was best grasped by imagining a
prepolitical "state of nature." The foundations of liberal politics are
not to be found in the realm of metaphysics. Rather, they consist of
hard-boiled observations of how ambition and the love of luxury,
diffidence and the love of glory, and the limitations of human reason
will always give rise to harsh disputes about justice. The foundation
of liberalism is a sober admission that the political animal is by nature
very unruly and cannot be self-governing without the outside help
of tricky and artful devices.

The democrats, as we have seen, claim that liberal political forms
are no longer necessary since technology, having reduced the sting
of scarcity, has softened the controversies of political life. In the age
of mature technology, the tensions held latent in liberal politics can be
brought to the fore and made the objects of genuine and democratic
deliberation. Now the democrats know well that the scarcity softened
by technology is not *simply* material. They know that while human
life must be preoccupied with material production, the end of life is
not merely the satisfaction of material needs. Rather, material production is motivated ultimately by the attraction of noble and beautiful
things, a fact that distinguishes human life from all forms below it.
And since the noble and the beautiful things are in essence political,
the democrats agree with Aristotle's view that physical necessity, or
productive mere life, entails politics, which concerns the good life.[23]

This "Aristotelianism" becomes clear when we consider the democratic (and Heideggerian) critique of bourgeois culture: the problem

23. In an earlier version of this essay I referred to these goods in more abstract terms as
self-sufficient ends in themselves. Mark Blitz convinces me that the more concrete terms
"beautiful and noble things" are much better.

with technology under the sway of totalizing metaphysics is the domination of life by cybernetics, utilitarianism, and instrumental reason, within whose scope there is nothing of human significance outside the range of production, reproduction, and consumption. Technological liberalism flattens the beautiful and noble things because it conceives of every human end as merely private and relative—good not in itself but only insofar as it is chosen by an autonomous self. Thus every end is the object of choice and *nothing* more, and every end is commensurable with every other. No end is a genuine root or anchor that gives moral depth and genuine, fixed character to a way of life. In such a world, all ends have in fact the character of mere means; there is nothing that money cannot buy and thus nothing that cannot be a medium of exchange and so merely useful for something else. Technological liberalism obliterates the possibility of genuine human ends; it thus "manages" political life by denaturing the soul.[24]

This democratic critique gives rise to questions. Why does liberalism so deform the soul? What liberal assumption explains this supposed propensity for self-mutilation? The answer given by the democrats: the part of Aristotle that liberals retain despite having rejected Aristotle as a whole. This is the view that the noble and the beautiful things are natural. Technological liberalism denies that such goods exist at all (there is no summum bonum, says Hobbes) because it can see no alternative to their being as Aristotle described them: natural, beyond the power of productive art, and thus scarce. As the state of nature reveals, the human propensity to believe in such goods causes murderous competition among people about the meaning of justice. If productive life continues to be motivated by the attraction of the noble and the beautiful things, then it will be harshly controversial no matter how productive that life might become. And so moved by the desire for mastery and control, the liberal simply denies their very existence and designs the technological world that keeps at bay the ever-lurking threat of naturalist superstition. For the democrats, the real error of technological liberalism is its own covert naturalism regarding the noble and the beautiful things; its view is that if they

24. As will become evident in the course of the discussion, I am suggesting that the problem of technology points beyond Kant's attempt to discover human dignity in the dictates of the moral law as opposed to natural teleology. This essay does not address that issue, however. It is much too long a story to be considered here. But of course in a more comprehensive account of our topic one would have to consider the Kantian aspects of liberalism and of the criticism of liberalism and the tension between Heidegger and Kant.

did exist they would be natural and could not be made, and that they must therefore be denied.

The democrats object to the liberal obliteration of the noble and beautiful things. But they are not therefore Aristotelians. Unlike their liberal antagonists, the democrats reject *completely* Aristotle's naturalism. They maintain that in our technological age nothing of political relevance—including a noble and beautiful end—is so scarce that its distribution would be the source of deadly conflict. How are these precious goods in sufficient supply? By being understood as they really are: as the products of aesthetic creativity. In the mature technological age, we can give up metaphysical naturalism properly and for good, something the old-fashioned liberal does not accomplish despite claims to the contrary. The noble and the beautiful things that Aristotle thought were natural, and that the liberal thinks do not exist, are actually the products of collective, aesthetic making. Arising from political creativity, they attain their status as genuine, life-forming ends.

Since it is aesthetic, there is no natural limit to the products of political creativity and no natural resistance in the "material" on which it works. It is, then, perfectly free as regards its ends and its easy means. Just like modern science, it does not have to submit to the recalcitrance of nature.[25] This constructive freedom gives its products—practical ends—the self-sufficiency worthy of humane longing and at the same time guarantees that there is always enough of them to go around. There is thus no essential split, no natural conflict, between various ways of life, such as once was thought to exist between the rich and the poor or the noble and the base. Nature does not make us equal, but material technology and political creativity can make us both equal *and* fully human. This is, of course, an unabashedly technological argument. From it we learn that the real problem of liberalism is its failure to grasp the gift of mature technology: not just material plenty but the final destruction of metaphysical naturalism. Residual naturalism is the unconscious foundation of the view that life is competitive and harsh and that politics must always take this fact into account. It likewise grounds the belief that technological rationality (and its attendant consumerism) is the only alternative to murderous partisan strife. But genuine democratic life will be

25. In Baconian science, we must submit to nature only in the process of conquering it. And for Kant, we must first see how science governs the phenomenal world in order to reveal the dualism on which genuine human freedom depends.

possible now that technology has freed us of this naturalism. Technology discloses the artificial character of the spiritual goods in the light of which common life can be experienced as a meaningful whole.

Now we must not be so impressed by the charm of this argument, or so daunted by the complexity of the theoretical issues at stake, that we hesitate to ask some simple and straightforward questions. And ask we must, since liberals are themselves inclined to a twofold trust in technology to dampen the fires of political life. On the one hand, liberals tend to think that we cannot live peacefully without an ever-expanding standard of living. And on the other hand, they tend to think that with abundance and the proper political arrangements—the separation of politics and civil society, representation, constitutionalism, and so forth—conflicts about justice and the good life can simply be made to disappear. There is a tension between the liberals' pessimistic foundationalism and their optimistic faith in Locke's suggestion that the poorest man in (technological) England will be grateful for being richer than the kings of the native Americans.[26] It could even be argued that the democrats simply radicalize and claim to fulfill the latter tendency to think that technology can render politics harmless. The democrats' hope to revive the political realm is not that far removed from the liberals' inclination to think that politics can be reduced to the techniques of administration. It is not an accident that the democratic point of view emerges along with the rise of liberalism to world responsibility and dominance.[27]

What questions must we ask? I think the first is whether the democrats exaggerate the distinction between material and spiritual goods. We should ask if the production of spiritual goods can draw forth material production and at the same time be free and unconstrained. Perhaps there is an unavoidable necessity for the pressing urgency of material production to contaminate the leisurely playfulness of politico-aesthetic creativity? If there is, then perhaps the good life, political as it is, will be much more contentious and fragile—and less obviously democratic—than the democrats contend.

Consider the example of Socrates in the *Apology*. It was his unstint-

26. John Locke, *Two Treatises of Government*, ed. Peter Laslett (New York: New American Library, 1965), pp. 338–39 (Treatise II, para. 41).

27. Until the time of the American Revolution and the Founding of the nation, no one seriously considered democracy a viable form of government, and only after World War II—under the influence of Hannah Arendt but during the heyday of liberal democracy—did anyone think of Aristotle as a "participatory democrat."

ing questioning of the claims to practical wisdom of the politicians, poets, and artisans that got him into trouble. By his own account of the matter, the artisans were the only ones who actually knew some-thing noble and good. Socrates, however, tells us that even they erred in taking their partial knowledge as evidence of their grasp of the greatest things, which we might identify as the just, the noble, and the good in themselves.[28] It was the democracy, the regime closest to the interests of the artisans, that found Socrates guilty and demanded his life. The reason for his condemnation seems clear enough. He appeared to the Athenians to have been an atheist and a sophist, at least according to the charges brought against him, and even at his trial he admitted his habit of exposing the ignorance of all (including his democratic judges) who thought they knew anything about the most important matters in life. In addition, after spending most of his time with rich young men, including the turncoat Alcibiades, Socrates found himself under a democratic regime that had just freed itself of oligarchic rule. Yet there is another, less obvious reason, which is particularly instructive for our question.

Plato thought, as did Aristotle after him, that human beings are moved to material production for the sake of the good life, that mere life is the means to the noble and beautiful things. But he also under-stood the following: just what the beautiful and noble things are is not self-evident, and since no one does anything other than what they think to be good, there is a powerful tendency for human beings to identify their own activities with the (rare) noble and good things. There is thus a powerful tendency for every art or practical activity to see itself as the self-sufficient end of the order of the arts and activities, to which every other art and activity is subordinate. Hence the certainty of the politicians, poets, and artisans, respectively, that *they* had knowledge of the beautiful and the noble (and, therefore, the just) and hence the partisanship of such claims to knowledge.

But despite their rancorous differences, they are all alike in com-parison to Socrates. How so? By comparison with them all, Socrates did nothing to contribute to practical life as they see it. He did not work; he did not take part in politics or take war seriously; he did not ask questions that got anywhere, but only those that provoked new ones; he seemed like a woman who did not give birth to children. (No wonder he was accused of corrupting the young.) The difference

28. Plato, *Apology* 22A-24C.

between Socrates and the partisans was that their lives were *useful* and his was not. This fact discloses the common error committed by the partisans and exposed by Socrates: they confuse their own useful activities with the noble and good things that are, as ends of the order of arts, not useful for anything. Whatever is practical is useful; whatever is noble and good is not. And yet the example of Socrates reveals that what is practical includes much that is far less obviously useful than mere manual labor or craftsmanship—it includes such things as virtuous deeds, pious behavior, poetry, and so forth. Everything noble and good, except for Socrates' philosophic activity, turns out to be in some way useful. And well they should. For if all were like Socrates, there would be no city at all.

Productive life—understood in the most expansive way to include all practically responsible activity—is thus moved by the confusion of the necessary (the useful means) and the self-sufficient (the noble and the beautiful end). This fact accounts for the partisanship of public life. Every responsible activity, however high or low or material and spiritual, can take itself to be the self-sufficient end of the social order. But since responsibility entails serviceability, every such activity makes a doubtful claim to being self-sufficient. Thus every claim (moral virtue, piety, military strategy, craftsmanship, poetry, or moneymaking) has to press its case against others who would deny it. But Socrates proved to be more obnoxious to every partisan than even their partisan enemies. For he appeared to them not only as impious and allied with a party they disliked but also as lazy and indifferent to the goods that for all of them made life worth living.

In the final analysis, Socrates' endless questioning conflicted with the urgency of practical life, with the seriousness of all the arts from the merely productive to the political and poetic. The leisure of Socrates' questioning of the noble and the beautiful things (including the gods) was at least as controversial as any opinion he might have suggested or criticized; for that questioning, pursued as a way of life, scorned every useful activity, whether concerned with material production or the performance of virtuous deeds or common deliberation. It is easy to overlook the simplest reason for Socrates' condemnation by the Athenians: the ineradicable tension between the production of useful things and deeds and the life that produces neither. In this respect Socrates' life stood for the paradox of human productivity: that for the sake of its own motion, production must despise the uselessness of the self-sufficient ends for which it works.

Since it assimilates those ends to the means it generates, the order of productive arts is always susceptible to partisan violence and resistant to reason and the life of the mind. Socrates was not just the victim of one particular regime. Rather, he succumbed to the city as such—as the place where things are made and deeds are done.

Has this paradox disappeared in our technological times? Is it true that aesthetic creativity can motivate production and yet not absorb the angry partisanship of those who take responsibility for urgent needs?[29] This question is pretty deep, but we can suggest an answer by remarking on just how much hard, serious work is required to maintain a technological society, and how serious and worklike is leisure in such a society.[30] One of the great errors of technological utopianism is the belief that an advanced technological society will no longer have to think of hard work as the most urgent and most dignified aspect of human life. No less now than in ancient times, the model of Socratic leisure would condemn the lives of the responsible. The example of Socrates would threaten the material foundations of a technological society, just as it threatened the simple Athenian artisans.

If it is not to stand as a reproach of responsibility, aesthetic creativity will perforce absorb responsibility's practical urgency and rancorous partisanship. There will be a powerful and necessary tendency for any one form of creative leisure to see itself as so important *as the end of material production* that it will not brook alternative forms tending to press the same claim. The activities of free creativity will become as controversial as were the claims of Socrates' constrained and busy rivals, the politicians, poets, and artisans. Thus in the democrats' projected order we should expect the most serious conflict between, say, hunting and photography, boxing and ballet, exotic dancing and square variety, misogynic rock 'n' roll and feministic folksinging.[31] And if we could imagine a world in which leisure and work could be wholly separated, a world where all production is done by robots sufficiently adaptive and inventive to operate completely on their own, matters would not really change. For then the robots would

29. Cf. Aristotle, *Politics* 1253b23–1254a9.

30. As any trip to Disneyland will prove. Heidegger never saw the Magic Kingdom or stood on line, cowlike and sweating, for the "Small World" ride. Thus he could not have imagined fully the industrialization of life and leisure in the age of technology.

31. It is easy to forget that we do not see such conflicts today because of the effects of liberal political forms.

be sufficiently human[32] to raise urgent, contentious questions about
their exploitation by and equality with their masters. It is doubtful
that the supposedly free, aesthetic creation of ends could both motivate production and transcend the urgency of production. If they are
to move production, such ends would have to be construed as absolutely serious, and would thus turn out to be seriously controversial.

The democrats are technological utopians because they presume
too clear a distinction between the goods of the body and the goods
of the soul, believing that when the former are abundant, the latter,
also plentiful, will not be the objects of serious political dispute. But
for the very sake of production as such, the goods of the soul must be
confused with those of the body and take on the urgency of material
necessity. They will be as pressing a source of controversy as are the
basic needs in less affluent times. Thus, even if the spiritual goods
are thought of as artificial and abundant, they will in practice be
pursued as if they were natural and rare. And the argument cuts the
other way as well: if production is necessarily moved by confusing
the good and the useful, it will always be possible to turn to the
merely useful things, as we do in practice to the goods of the soul—
that is, as if they are in relatively short supply. It will always be
possible (indeed it will be necessary) for some to claim—on the basis
of their superior dignity—a right to more material goods than others,
even when more can always be produced.

But the argument thus far raises another question. For the democrat could object at this point that while technological utopia really
is "no place," the contentiousness of productive life can at least be
softened by the counterthrust of aestheticism. The truth about the
noble and the beautiful things *is* that they are in ample supply because they are freely made. And this truth can be brought to bear on
practical life even though it cannot completely determine that life.
Thus the urgency of production will make its mark on political practice, but not so powerfully as to make politics too dangerous a field
on which to play. Politics will be like play, but yet serious enough to
be the most important, the most ennobling and liberating human
activity. Is this not the more realistic position we originally ascribed
to the democrats?[33]

Again, we are confronted with hard questions about the noble, the
good, and the just. One could without exaggeration say that we are

32. Certainly far more so than the animals to whom some liberals would now grant rights.
33. See p. 260 above.

thrust right back to Platonic political philosophy. But again we might get our bearings here by staying down to earth and not jumping into the arcane matters of Western metaphysics and foundationalism. We should also avoid the rationalistic dogmatism that does tend to cover up the phenomena of practical life—such things as utter abandonment to God and the call of faith, the beauty of courageous sacrifice, the thrilling danger of political crisis, the world-shattering riskiness of art.

If we do, then a curious thing happens. The aestheticism of the democrats begins to look very much like the cybernetic, stultifying, and bourgeois utilitarianism they so despise in their image of liberal morality, politics, and culture. The democrats aim to resurrect the noble and the beautiful things and to make them politically relevant, but not to the extent of requiring real sacrifice or pain. They oppose the call of any noble or beautiful thing whose power and depth would have the effect of hurting or excluding anyone. Indeed, for the democrats liberalism is just as bad for its tendency to harm and exclude as it is for its soul-deadening philistinism.

In the democratic scheme, all people must be not just free to be creative (any liberal can accept this), but actually as creative as they wish—and in public as much as in private. Of course this would be possible only if the supply of noble and beautiful things is always large enough to go around, as it is since they can be made at will. These goods would thus be equal as mere creations, and they could not be otherwise without limiting the aspirations of some: no one's creation can make a greater claim on material resources or respect than anyone else's. It goes without saying that such a scheme could not accommodate a person's unceasing (that is, really serious) creativity, or tolerate great and commanding creations, or allow for the longing for a community based on service to the one true god. Moreover, a *created* self-sufficient end is simply an oxymoron. How could such an end command our longing when it is dependent on its maker? The concern for inclusion and avoiding hurt calls for the bourgeois attraction to long life, commodious living, and easygoing relativism. It does not call us to fate, to the commanding sublimity of one's community, to the rare and independent beauty of noble deeds, to the love of danger, or to self-forgetting faith. It is thus not surprising that for all their desire to *épater le bourgeois* the participatory democrats are professors who, after a long day in the classroom slaving creatively over Nietzsche, go home to jog, mow their suburban lawns, or—at their most daring—don the mantle of well-known cosmopolitan intellectual. In opening up the question of the noble and the beautiful things, the democrats answer in even softer, flatter, and more technological tones than does the old-fashioned liberal.

Yet the democratic critique reveals much about the character of technological liberalism. It does tend to stifle deep longings of the soul. But as a danger that saves, this stifling paradoxically brings such longings to the fore—our very awareness of the problem of technology attests to them and the phenomena that attract them. Technological liberalism is enormously resourceful, however, refashioning these longings after its own image, as we see demonstrated in the case of the democrats. Even so, technology *is* ultimately a danger that saves, no matter how resourceful liberalism might be in opposing its gift. Short of the complete denaturing of the soul, there can be no complete obliteration of the noble and the beautiful things. The harder technology presses, the more we can expect to sense their absence. Their call is always possible, and so it is always possible that there will be genuine listening. We can thus expect no simple cures for the dissatisfactions of the liberal soul.

Two tentative conclusions follow. First, the darkest possibilities of political life, those having to do with the call of the gods, the authority of the rooted locale, and the lure of the uncanny and the strange, may not be so easily overcome as the democrats and technological utopians would have us think. The greater the thrust of technology, the greater may be the counterthrust against its nihilism. If so, then commonplace partisanship may be more dangerous than it seems, tied as it is to greater and more turbulent depths. All the more reason, then, to preserve the forms of liberal politics with an eye to managing such forces, and to consider that our liberal institutions may, in fact, be managing them as we speak. We should not be so impatient with the weaknesses of liberalism that we fail to appreciate what we gain from it.

Second, we should not assume that the first liberals were unconscious, self-contradicting, metaphysical naturalists. They might have been fully aware of the premises of their sober foundationalism, basing it on a well-honed understanding of politics—especially on a subtle account of the noble and the beautiful things and their relation to productive and political life. We must resist the temptation to think that the first liberals were in the grip of dogmatic utilitarianism. Likewise, we must not assume that liberal politics is incapable of attending to the soul's deeper longings. Liberals can and certainly have allowed for the virtuous and reasonable cultivation of these longings. What nobler and more taxing task can be imagined, what greater challenge to statesmanship and military virtue, than the liberal effort to save the world from the dark night of fascism and communism? What more thrilling and exacting danger than having managed nuclear deterrence, which approximates antiquity in having the soldier and the statesman share the same risks? What grander common project than the conquest of space? In a world where most suffer at the hands of poverty, despotism, and malevolent clerics, what more inspiring sight than a people who can govern themselves?

Finally, liberals need not focus only on their capacity for the practical virtues. It is possible that philosophy has never been more urgent than in our liberal, technological times. What greater challenges are there to thought than the tensions between the good and the just, and between reason and faith, which grow so readily out of the rough and ready compromises of liberal politics and the concomitant project for conquering nature? What greater challenge is there to thought than the Heideggerian argument about technology, which raises again the question of Being and arises in reaction to our liberal, technological time? Indeed, liberals would do well to admit that with respect to the noble and the beautiful things—and with respect to the highest possibilities of thought—the liberal political order is where most of the action has been.

Much of what has just been suggested about the foundations of political life informs against the progressive view that the liberal consensus is now so complete—so much a part of inevitable moral progress—that it no longer propels liberals on the impossible quest for metaphysical truth. We have suggested that the real foundations of liberalism are more anthropological than metaphysical, more based on political philosophy than on some demonstrable, theoretical account of Being and the natural whole (which is not to say that early liberals did not think seriously about such matters). Take Hobbes and Locke, for example. Surely their philosophical skepticism is a far cry from some metaphysical foundationalism. In fact, they argued against counting on philosophy and metaphysics for political guidance. On the one hand, they thought there was just too much disagreement in philosophy, and on the other hand, they thought that metaphysics gave too much encouragement to religious enthusiasm. In this respect, they anticipated the progressive case against linking politics to a demonstrable, theoretical account of the natural whole.[34]

But then a simple question arises: Why do the progressives claim to have discovered what in some respect has always been evident, not to say obvious, in liberal thought? Or put another way, why do the progressives see liberalism only through the eyes of Kant's critique of metaphysics and his doctrine of the absolute moral law?[35] Perhaps there is just an ax being ground in the excited refutation of a foundationalist straw man. Is there not something suspiciously "foundationalist" about the attack on foundationalism—does not this attack assume that getting things straight about politics requires the certain foundation of having the correct theoretical stance toward foundations: that there aren't any? Is this not more theoretically foundational than anything suggested by the early liberals?

Again, these questions go far beyond the limits of our available space. But we can at least put our finger on the basic problem, which has to do with the relation of thought and politics. For the progressives there are only two possibilities: the mistaken view that a rational political arrangement—liberalism—depends on the demonstrable truth about nature and the whole, revealed by metaphysics; or the correct view that metaphysics is impossible and that politics is and always was prior to thought. The correct view comes to light only with the course of progress, which the progressives interpret in the light of Heidegger's account of Western rationalism and its culmination in technological nihilism. Reason first claimed priority over politics with the advent of Socrates, the first metaphysician, who assumed that a term in his final vocabulary has a real essence or eternal and knowable *eidos*.[36] So grounded, the terms of Socrates' final vocabulary were established as the principles for determining justice and the best political order. Socratic rationalism (or foundationalism) continued unabated through the course of Western history, culminating in the Enlightenment, liberalism, and the project of modern science. It then unraveled under the onslaught of German Idealism, Romanticism, and the critique of technology. Only now, at the end of this long saga, can we see for the better that the relation of thought

34. Thus for similar reasons they were also concerned about what gets taught in the universities.
35. Rorty, *Contingency, Irony, and Solidarity,* pp. 4, 32.
36. Ibid., p. 74.

and politics is just the opposite of the view imposed on us by Socrates.

We cannot begin to check out this story in any detail. But as with our examination of the democrats, the example of Socrates turns out to be relevant to the question of liberal politics. For the progressives, liberalism comes into its own only when the dead weight of Socrates is finally lifted from our backs. There are some problems with this story, however, especially as regards its beginning. First, it ignores the fact that Socrates' discussions of the *eide* of things always transpire in the dramatic context of a dialogue—and thus are always determined in some way by the prejudices and character of the interlocutor and by some urgent practical and political situation. Second, it ignores the fact that perhaps the most explicitly political of such discussions, the one in the *Republic*, is in large measure both comic and mythic. Actually, Plato's dialogues are first and foremost hermeneutical, not epistemological and metaphysical, which is why they always end in doubt, not wisdom. That Socrates offers no proof of the *eide* does not mean that he simply assumes them. About their existence he assumed nothing; there is no assumption that Socrates does not doubt. And yet while he does not assume the existence of the *eide*, as Rorty asserts, if we are not ourselves to distort Socrates' questioning, we have to admit that he does take them seriously—that we cannot imagine Plato's Socrates apart from the *question* of the eternal, or natural, "whats" of things and the relation of this question to practical moral life and politics.[37]

Socrates was like Rorty's description of him, but not the same. So who was he? We can find out by considering the odd fact that Socrates—supposedly the opposite of the ironic liberal—was himself ironic. He claimed only to have knowledge of his own ignorance, which required that he know quite a lot. And as we noted earlier, his questioning never seemed to get anywhere, at least in the eyes of the responsible Athenians. So how does Socratic irony differ from

---

37. In the effort to keep things simple, I here avoid the question of just what the "whats" might be. It will be sufficient to say that Heidegger's powerful attack on the epistemologically privileged position sought by the metaphysical tradition—a position outside the finite assumptions of some concrete, particular world—does not by itself uproot the question of the being of nature. Rather, I would say, it liberates this question from the dogmatism and moralism of Enlightenment rationalism, which still lives secretly in progressive postrationalism. Heidegger helps us to reconsider the relations among thinking, knowing, Being, and nature, without equating them. It may well be that liberalism depends on our openness to the structure of a natural whole, but that this openness can never produce the certainty demanded by every partisan, including the progressive.

progressive irony? In two separate but related respects: first, Socrates sees the need for politically useful lies; second, he takes seriously and on their own grounds the political opinions he investigates, as if they might be either true or false. To use Nietzsche's terms, Socrates' irony combined playfulness with the spirit of gravity. The reason for this is that Socrates avoids any dogmatic assumption about the whole, including the assumption that it is the product of art. Consequently, he was attuned to all the phenomena of practical life, including those serious ones that can be seen fully for what they are only in their openness to something like nature or the *eide* or the gods—that is, such things as the noble, the just, and the sacred.

But the fact that Socrates was attuned to these phenomena does not mean that he thought he could answer the questions that they provoke. For him, the life of questioning was not the same as the attainment of wisdom, at least not wisdom as it would be understood and required by those responsible for material production and political life. They claim to know what the noble and the just are and to know about the gods. Socrates did not. And yet for him these claims set the terms of the philosophic life: without the question of the noble, the just, and the gods, there would be no wonder at all. Thus Socrates knew three things well: first, how endangered his philosophic activity was by the inescapable dogmatism of the responsible; second, how imperiled thought was by the political market for dogmatism—how easy it would be for thought to serve the responsible (as was evident, for example, in the sophists); and third, that the life of the mind is inescapably framed by the perplexities arising from human responsibility. Were all material goods to come from heaven, there would be no life of the mind.

Socrates was playful because he could not be responsible without succumbing to dogmatism. He was serious because he knew that the questions posed by the responsible are coeval with political life, that politics gives life and horizon to the mind, and that politics is at the same time hostile to the fullest thoughtfulness and wakefulness. Contrast this with the progressive ironist, who is only playful and never genuinely serious. For Socrates, such unadulterated playfulness would reveal a blindness to the fractured character of political life and to the paradoxical relation between thought and responsibility. As such, it would be more like the partisan's claim to know—like the partisan's typical belief that there is an immanent common good—than it is like the activity of being most doubtful and awake. Confronted by the progressive's unmitigated playfulness, Socrates would no doubt probe so as to discover the partisan and moralistic enthusiasm lurking behind the ironic and supposedly privatized smile.

Some evidence in this direction is provided by a look at how the progressives describe the inevitable liberal consensus. According to Rorty, it consists in the distinction between public and private life and the abhorrence of cruelty. But it also includes the following: democratic socialism, contempt for religion, hostility to constitutionally managed political competition, hostility to the institution of private property, and the conviction that Ronald Reagan's "manipulators," just like the Moscow "nomenklatura," were "thugs."[38] A Socratic grilling would indeed have borne fruit. One is simply amazed to see how partisan, not to say repulsive, such views must appear to most nonphilosophic liberals, even those whom Rorty would classify as ironized. The issue here is not that the progressives dogmatically assume a truth to which by their own position they are not entitled.[39] Rather, the point is that liberal ironism obscures the existence and full character of the problems generated by such views. The progressive—the ironic doubter of all moralists—thus exhibits the most common trait of cantankerous moralism: the belief that the context in which practice transpires is so smooth and well woven that we face no real problems or perplexities and no danger from partisan zeal.[40]

38. Rorty, *Contingency, Irony, and Solidarity*, p. 196; "Thugs and Theorists," *Political Theory* 15 (Nov. 1987), 565–67. He argues this in the otherwise sensible context of explaining the difference between the United States and the Soviet Union.

39. Again, the easy charge of self-refutation has to be resisted. First, it is not true in the case of Rorty, or Heidegger, or any serious historicist. Rorty may be committed to a view of the whole, but that does not mean that it must be "metaphysical" in the way that Rorty understands the term. Hence he is not guilty of some simpleminded self-contradiction. Second, the self-refutation ploy just prevents our taking seriously what both Rorty and Heidegger wish to understand: the founded and finite character of the worlds within which thought and questioning occur. See note 18.

40. Rorty, as a moral skeptic who serves moralism, makes underlying ontological assumptions that are ultimately metaphysical in the following, negative sense: for him the historical overcoming of epistemology and metaphysics discloses a whole that must be artificial since it does not measure up to the standard of what natural being *would* be if such being existed— demonstrable knowability. My argument here does not beg the question of whether or not there are genuine problems about the good and the just, let alone that such problems can be resolved demonstrably. Rather, it suggests that an ontology that admits such problems is the basis for the fullest wakefulness and attention to the contours of practical experience. The "proof" of this ontology would consist in showing a life of such wakefulness to be possible, in picturing it, and in the extent to which such a picture would light up the most immediate and concrete aspects of political life. From such a "proof" we could not derive a demonstrable morality, which is not to deny that we could extract from it a significant political prudence. Plato's dialogues come to mind.

If there is anything to this argument, then the danger of progressivism is that it generates an extreme political naiveté and an exaggerated faith in the power of thought to penetrate political life. Since
metaphysics is the real source of our ills, its demise with the progress
of technology will make politics a form of play. One holding such a
view could easily believe in the inevitability of an overpowering liberal
consensus. From a more Socratic point of view, however, such a consensus is always fragile. Consider the following: liberalism stands or
falls with an expanding standard of living, which recent experience
shows depends on capitalism (socialism, we now know, fetters the
means of production). The rising standard of living does away with
crushing inequalities (as can be found in an aristocracy or traditional
society, for example). But this general equality is purchased by an
economic system that generates smaller and less crushing, but ineradicable, inequalities. As Tocqueville has shown, in democratic and
egalitarian societies the slightest inequalities matter very greatly,
which in turn makes the longings for equality at once insatiable and
irritable.[41] There is every reason to think, therefore, that a liberal,
egalitarian consensus will be forever threatened by the age-old cleavage between the rich and the poor, however better off all may be
compared with earlier times. This cleavage generates the most pressing claims about the good and the just, claims that cannot simply be
ironized and that may require the most subtle management, both
political and moral, for their practical resolution. Again, the issue is
not that we need a demonstrable morality, but that we must not let
complacent, progressive moralism take liberalism for granted.

The early liberals thought that a reasonable political order could
be built only with a clear understanding of the limits of reason's power
to inform individual behavior and political life. This was the meaning
of their sober foundationalism and their reliance on the artful device,
the clever deceit, the benign manipulation of the passions (Machiavelli was their first teacher, after all). In this crucial respect, the early
liberals have more in common with Socrates than they do with their
own late progeny, the progressives. This fact is remarkable, since the
liberals turned against the practical moral and political teachings of
the ancients, who certainly would have disapproved of the most important aspect of the liberal program: the technological project for the
conquest of nature. In fact, the latter project inclines liberals toward

41. Alexis de Tocqueville, *Democracy in America,* trans. George Lawrence (Garden City,
N.Y.: Doubleday, 1969), pp. 535–38.

progressivism for all the reasons we have surveyed. Liberals do tend to expect that some hidden hand of progress—always linked to technology—is transporting them to a wholly transparent world in which all we need to do is acquire, become creative, or relax in smug irony in the playful realm of civil society.

Thus as the liberal project matures and gathers its technological steam, we are forced almost paradoxically to think about its kinship with thought that is quite distant from its moral and intellectual shores. There is a way in which liberal principles do not stand entirely on their own, even and especially when they seem to be triumphant. Perhaps the first liberals would tell us as much if we return to them without the almost canonical picture of the "canon" of liberal thought drawn by the progressives and the democrats. We might begin by rethinking the apparent novelty, dogmatism, and foundationalism of the first proponents of the scientific conquest of nature (such as Machiavelli and Bacon), and of the first liberals who were so profoundly influenced by them (such as Hobbes and Locke). We might get a clearer picture of liberalism from them than we can from many who speak for it today. These earlier thinkers actually reflected more deeply on technology than we do, in large part because they could not take it for granted. Consequently, they thought more seriously about the differences among art, politics, and thought. We would do well to follow their lead.

Both the progressive and democratic interpretations of Heidegger inflate the importance and independence of thought in practical affairs. They tell us that epistemology and metaphysics—bad ideas— are the remaining impediments to the thoughtful construction of meaningful common life. For the democrats, the common good is secured by thought. For the progressives, the common good is secured for thought. And for both, thought is merely the freewheeling vehicle of the aesthetic imagination. The argument of this essay suggests that we might well think twice about the power of material or intellectual forces to release thought from the grip of material necessity. We learned this lesson in part from Plato and Aristotle. We suspect that it was well understood by the first liberals and is at least implied in the mind-set of the old-fashioned liberal fogies of our day. And yet in a progressive and democratic climate, when fancy theory and common opinion mingle so readily, thinking about liberalism will be ever more important in establishing reasonable liberal bearings.

In the end, the fate of liberalism may depend (among other things, of course) on our finding within it an openness to thought prior to its own technological horizon. The danger of our time is that technology will promise a wholly changed world when there is really not much new under the sun.

Suggestions for Further Reading

Arendt, Hannah. *The Human Condition.* New York: Viking, 1958.

Barber, Benjamin. *Strong Democracy: Participatory Politics for a New Age.* Berkeley: University of California Press, 1984.

Borgmann, Albert. *Technology and the Character of Contemporary Life.* Chicago: University of Chicago Press, 1984.

Caton, Hiram. *The Politics of Progress: The Origins and Development of the Commercial Republic, 1600–1835.* Gainesville: University of Florida Press, 1988.

Ellul, Jacques. *The Technological Society,* trans. John Wilkinson. New York: Knopf, 1964.

Ezrahi, Yaron. *The Descent of Icarus: Science and the Transformation of Contemporary Democracy.* Cambridge: Harvard University Press, 1990.

Habermas, Jürgen. *Toward a Rational Society: Student Protest, Science, and Politics,* trans. Jeremy J. Shapiro. Boston: Beacon Press, 1970.

Jonas, Hans. *The Imperative of Responsibility: In Search of an Ethics for the Technological Age.* Chicago: University of Chicago Press, 1984.

Marcuse, Herbert. *One-Dimensional Man.* Boston: Beacon Press, 1964.

Schürmann, Reiner. *Heidegger on Being and Acting: From Principles to Anarchy.* Bloomington: Indiana University Press, 1987.

Weinberger, Jerry. *Science, Faith, and Politics: Francis Bacon and the Utopian Roots of the Modern Age.* Ithaca: Cornell University Press, 1985.

Winner, Langdon. *The Whale and the Reactor: A Search for Limits in the Technological Age.* Chicago: University of Chicago Press, 1986.

Wolin, Sheldon. *Politics and Vision.* Boston: Little, Brown, 1960.

Zimmerman, Michael. *Heidegger's Confrontation with Modernity: Technology, Politics, and Art.* Bloomington: Indiana University Press, 1990.

12 The Problem with the "Problem of Technology"

ARTHUR M. MELZER

It is not an exaggeration to say that the question of technique has now become that of the destiny of man and of his culture in general.
—Nicholas Berdyaev (1934)

Thinking is called into existence by problems. The thought not only of individuals but of whole eras thus tends to organize around some prevailing conception of the deepest question. In our own age this catalytic, crystallizing question seems increasingly to center on technology. All the great issues of morality and politics, formerly addressed on their own or in a religious context, are now placed in the larger context of technology, which suddenly seems the most powerful and pervasive force, the ultimate phenomenon.

The clearest evidence for this change is the remarkable convergence that has occurred over the last sixty years between theorists of the radical right and left: both now regard technology not merely as problematic but as our deepest problem. On what may be called the Romantic and existentialist right, there has, of course, been a long tradition of hostility to machine technology, perceived as destructive of man's higher aesthetic and spiritual nature. But in Heidegger, this tendency is deepened and placed at the center: technology, properly understood, is the most fundamental as well as the most problematic

287

characteristic of human existence in our age and, at least incipiently, of Western thought and Being since its Platonic origins.[1]

On the left, the dominant, Marxist view—which arose by repudiating an earlier tradition opposed to industrialization as inherently exploitative—attributed all exploitation to the mode rather than the means of production, to the bourgeois ownership of technology rather than to technology itself. By thus shifting all blame for injustice onto capitalism, Marx not only exculpated technology but was able to embrace it as the ultimate savior and liberator of humankind—a view that triumphed because it was more consistent with left-wing progressivism.[2] But in the last sixty years the thought of the Frankfurt School and its spin-offs in the new left have produced a veritable revolution on the left, so that what was formerly embraced as a savior has now been revealed as the devil itself. Today on the left, no less than on the right, technology and "instrumental rationality" are seen as the root phenomenon of modernity and the true source of all our evils.[3]

The emergence of technology not only as a problem but as the organizing problem of our times can also be seen in other, less theoretical phenomena. In contemporary popular culture, for example, what is the largest, weightiest, most "philosophical" question that it is possible to ask? Not whether we will be saved, not whether there will be a transfiguring political revolution, but whether we will destroy the planet. Especially for those of college age, that is the truly *serious* question that quiets the unruly crowd and seems to place everything else in its proper perspective.

To exaggerate only slightly, one could elaborate the apocalyptic claims implicit in this conception by speaking of "three ages of man." In the first, the *metaphysical age*, the question that organized human thinking concerned the individual's relation to the divine or transcendent order. In the second, the *humanistic* period, the issues of humanity's worldly security and comfort, and the problem of man's

---

1. Martin Heidegger, "The Question concerning Technology," in *The Question concerning Technology and Other Essays*, trans. William Lovitt (New York: Harper & Row, 1977) and *Nietzsche*, vol. 4, *Nihilism*, trans. David Krell (San Francisco: Harper & Row, 1987). See also Schürmann and Vattimo, Chapters 8 and 9, respectively, in the present volume.

2. Consider, for instance, Friedrich Engels, *Socialism, Utopian and Scientific*, trans. Edward Aveling (New York: International Publishers, 1935).

3. See, for example, Max Horkheimer, *Eclipse of Reason* (New York: Seabury, 1974), and Herbert Marcuse, *One-Dimensional Man* (Boston: Beacon Press, 1964). See also Wolin, Chapter 7, in this book.

exploitation of man (the "social question"), moved to the center. In our own age, the third, those problems have not disappeared, but the crucial issue has become humankind's relation to the rest of nature, understood now not as transcendent but as something vulnerable and subject to our control, for which we must take responsibility. This is the age of *the problem of technology.*

Certain as it is that the focus of human thinking has changed radically over time, that fact alone is obviously no proof of the validity of that change. The human diet has also varied over time, which does not prove that some foods do not nourish better than others. It is at least possible that the deepest questions of human existence are fixed—rooted in an unchanging human condition—however neglected or variously distorted they may be in different ages. Thus, in knowing ourselves to view the world from a new and altered perspective—that focused on the question of technology—it is still incumbent upon us to wonder how this change came about and to raise the question of its value or legitimacy. This is, of course, a monumental undertaking. What follows is merely an attempt to survey the terrain, some of it familiar and some new.

Our Uniquely Technological Age

It is an obvious fact that neither technology nor the question concerning its value is new to the modern period. Indeed, human life as such is so inseparable from technology that homo sapiens is commonly distinguished as "the tool-making animal," just as the stages of human civilization are differentiated in terms of the tools human beings have made. The pervasive effect of technology is visible even in the most primitive societies, especially if one considers not only simple tools and weapons but also ritual art, body painting and surgery, fermented drinks, and so forth. And no recent technological innovation has been so momentous in its consequences as the agricultural revolution.

As for the awareness of technological change as a problem, one might point to the fierce resistance of hunters and nomads to agriculture, which easily rivals the more recent hatred of peasants, artisans, and intellectuals for the machine. On a more reflective level, the stories of Babel and Prometheus suffice to show that some kind of "problem of technology" has been on human minds for a very long time.

And yet clearly there is something distinctive about the modern period. We not only have more technologies, but somehow we are more technological. One senses that the phenomenon of human artifice, which is in some sense universal, has come into its own so radically and pervades human existence so thoroughly that it now seems to form the horizon within which everything else finds its place. If one could clearly spell out the ways in which this is true, one might see why the problem of technology has taken center stage today, as well as begin to evaluate whether it has a right to that place. Thus we need to understand two interrelated things: how is modern technology unique and how is modernity uniquely technological?

I first present a brief survey of more or less familiar characterizations of technology and of its role in the modern age and then turn to the two factors that seem to me the most important as well as the least adequately understood—the modern belief in progress and disbelief in happiness.[4]

The Greater Extent of Modern Technology

The most obvious characteristic distinguishing modern technology from all earlier forms is simply that there is such a great deal *more* of it. Of all the objects with which one routinely comes into contact today, a remarkably high and growing percentage is artificial. These technological objects, of which there are more, are also more technological—being artifacts not merely in form, like an earthenware pot, but also in their very substance. "Plastic" denotes a new level of artificiality. Furthermore, they are *machines* and not merely tools, for even their motion or power is artificial.[5] And increasingly they are "smart" machines, replacing the power of both the human mind and the body.

They are further distinguished by the involvement of new, unanticipated physical forces or effects, such as nuclear energy or laser optics, which are wholly alien to our natural experience of the world. They also stand at a greater remove from our natural productive faculties, being the product not of human hands directly but of machines that

4. For a very useful survey of views on the phenomenon and problem of technology from a different point of view, see Carl Mitcham and Robert Mackey, *Philosophy and Technology: Readings in the Philosophical Problems of Technology* (New York: Free Press, 1972), pp. 1–30.

5. Hannah Arendt puts particular emphasis on this factor, since it forces the worker for the first time to submit to the alien motion and rhythm of the machine. See *The Human Condition* (Chicago: University of Chicago Press, 1958), pp. 144–53.

are the product of still other machines, and so forth. Finally, they are technological not merely as *means* for the provision of natural or at least accustomed goods but exist for the service of ever more artificial *ends*. The purpose of a television set, for example, cannot be fully understood except in terms of a world of needs as artificial as the television itself.

It is not merely artificiality, however, that distinguishes the products of modern technology, but also ever-increasing automaticity, uniformity, and disposability. These are the qualities that give rise to the unique sense of sterility, the existential disengagement and alienation from the world of objects, that seems to be a hallmark of modern experience.[6]

New Relations of Dependence

Two further characteristics of the modern world follow directly from the great extent of technological development. Our age lives in a condition of utter and unprecedented dependence upon technology. In the absence of the artificial world we have created for ourselves, very little of modern life would remain. For example, the prosperity that has made possible mass education and thus mass democracy would disappear. Indeed, perhaps 85 percent of the world's present population, which has increased almost eightfold since the start of the Industrial Revolution, would be insupportable.[7]

Just as striking, however, is the degree to which the rest of nature has become dependent on us. Because of the vast scale of technological development, it is now up to us whether other species will survive or disappear, whether the atmosphere and climate will continue unchanged, whether the planet will remain supportive of human and other forms of life. In all past ages, it was believed that, however great man's evil and foolishness, his sins only fall back on himself, leaving the larger order of nature to go its way serenely unchanged, a palpable symbol of man's pettiness and his need to conform to the order that transcends him. In fact, it still remains true, if no longer so palpable, than human beings have no power over the basic laws of

6. On this last point, see especially Albert Borgmann, *Technology and the Character of Contemporary Life: A Philosophical Inquiry* (Chicago: University of Chicago Press, 1984).

7. See José Ortega y Gasset, "Man the Technician," in *Toward a Philosophy of History* (New York: Norton, 1941), and *Revolt of the Masses*, trans. Anthony Kerrigan (Notre Dame, Ind.: University of Notre Dame Press, 1985).

nature or over the ultimate finitude of human life and the cosmic conditions that support it. Nevertheless, it is difficult to escape the conclusion that something of a revolution has occurred in the structure of the human condition as a result of modern technology. All future thinking about morality and the human good must include the new fact that human beings have largely succeeded in becoming the "masters and owners of nature," as Descartes had urged, and that this ownership brings with it new responsibilities. The classical virtue of moderation is as crucial today as it was in the past, but henceforth it will be based less on nature's overawing permanence than on its vulnerability, less on the smallness of human beings than on their excessive and dangerous technological power.[8]

The Greater Scope of Modern Technology

It is not only the extensive material development of the modern world that makes it uniquely technological, but also the spread of this phenomenon to other areas of life formerly untouched by it. Beyond physical instruments and machines lies something that may be called the technological "attitude," or "way of thinking," or even "posture toward Being": a nonspecific but generally utilitarian understanding of ends, a primary focus on means and power, the restriction of reason to instrumental rationality—the methodical pursuit of the one maximally efficient way of doing each thing—the faith in human self-reliance and control, the belief in the superiority of the artificial to the natural and of the mechanical to the human, and the view that everything man encounters in nature or history is only raw material that he is free to transform for his own purposes. As has been shown with particular force in the work of Jacques Ellul, this attitude, which used to be found in only a few narrowly circumscribed areas of life, now seems to dominate virtually all.[9]

8. On this theme, see the beautiful discussion by Hans Jonas in *Philosophical Essays* (Englewood Cliffs, N.J.: Prentice Hall, 1974), pp. 3–20. See also Andrew Goudie, *The Human Impact on the Natural Environment* (Oxford: Basil Blackwell, 1981); Carolyn Merchant, *The Death of Nature: Women, Ecology, and the Scientific Revolution* (New York: Harper & Row, 1980); and William McKibben, *The End of Nature* (New York: Random House, 1989). Although long oblivious of these threats, today we often seem to be in danger of overestimating the fragility of the earth.

9. See Jacques Ellul, "The Technological Order," in *The Technological Order*, ed. Carl Stover (Detroit: Wayne State University Press, 1963), and *The Technological Society*, trans. John Wilkinson (New York: Knopf, 1964); see also Langdon Winner, *The Whale and the Reactor: A Search for Limits in the Age of High Technology* (Chicago: University of Chicago Press, 1986), pp. 47–54. See also Introduction by Kass in this book.

The whole idea of "revolution," for example, which stands behind most modern political thinking, imitates man's conquest of physical nature. As Robespierre urges, "Everything has changed in the physical order; everything must change in the moral and political order. Half the revolution of the world is already accomplished; the other half must be achieved."[10] All liberal political systems in particular are decisively technological in theory and practice. For such thinkers as Locke, Montesquieu, and the American Founders, the state is seen as an artificial human invention, needed to correct the gross defects of nature, and as something fundamentally instrumental, a mere means for the provision of private ends. The citizens, as well, are conceived as "individuals," that is, independent human atoms subject to internal drives and external manipulation. The machine of state that organizes them is to be constructed not in imitation of some natural or divine pattern but in the most effective and efficient way. Thus great reliance is placed on constitutional mechanisms and techniques—separation of powers, checks and balances, representation, party government, federalism, "extending the sphere," the free market—rather than on human qualities like statesmanship or moderation and justice. As stated (perhaps overstated) in the title of Michael Kammen's book on the American Constitution, the goal was to build "A Machine That Would Run of Itself."[11]

Similarly, in the subpolitical realm of work and economics, we have created an instrumentally rationalized world of bureaucracies and assembly lines, of specialized people and standardized products. Even on the level of leisure and amusement, one sees a similar tendency, perhaps less planned but no less obvious, toward the triumph of technique. Indeed, it is difficult to think of any human activity, grand or trivial, for which there does not now exist a specialized method, therapy, or system, whether for eating or fighting, for dating or proselytizing, for relaxing or motivating, for making cars or making love. Time itself has been captured by clocks and schedules, space by urban grids and highways, value by dollars and cents. Gradually all alternative principles of experience and choice—piety, morality,

10. Maximilien Robespierre, "Report on Religious and Moral Ideas and on the National Festivals," in *Oeuvres complètes*, 10 vols. (Paris: E. Leroux, 1910–67), 10:444.

11. Michael G. Kammen, *A Machine That Would Run of Itself: The Constitution in American Culture* (New York: Knopf, 1986). See also McWilliams, Chapter 4 in this book.

aesthetics, custom, instinct—have been, to paraphrase Marx, drowned in the icy waters of technical calculation.

The New Scientific Basis

A still more distinctive and profound characteristic of contemporary technology is its intimate connection with modern theoretical science. In prior ages, each of the different arts was based on a relatively separate and self-contained body of practical know-how or crafts knowledge, acquired through trial and error and transmitted through tradition. Indeed, the practical arts were *necessarily* separate from science or theory, owing to the latter's motive—the idle, disinterested contemplation of what is—as well as to its object—the realm of changeless, eternal beings that cannot be other than they are. A revolution in human thinking was required to produce the situation that we now take for granted: the derivation of the crafts from natural philosophy.

The most obvious consequence of this revolution has been the production of remarkable new craft objects that are able to harness forces of nature unsuspected by the prescientific, artisanal world. A deeper if less visible consequence, however, is the fundamental transformation of our attitude toward these practical objects, which now carry with them all the splendor and dignity of "theory." Modern science constitutes the dominant theory of reality, the highest intellectual authority of our age, much as the Bible did in former times. Today technological objects have therefore taken on something of the character of sacred objects or divine gifts. They are no longer seen as merely the fruit of some particular craft or artisan but as a remarkable demonstration and consequence of our successful contact with the deepest ground of reality, a sort of message or gift from on high. In other words, prior to our age, there had never been a world in which the most ordinary, mundane objects were so directly and *convincingly* linked to the most ultimate principles. Our great veneration for high technology and its peculiar prestige in the world—its use, for instance, as the dominant measure of a society's level of civilization— would seem to derive less from the material comforts it provides than from the belief that it constitutes an indubitable revelation and local incarnation of the deepest forces of the universe.

The modern convergence of the arts and natural philosophy, however, involves the transformation of the latter no less than the former;

and this change is responsible, on a still deeper level, for the uniquely technological character of our age. At the dawn of the modern era, natural philosophy was transformed into modern science, which is, *in itself,* even prior to all practical applications, essentially technological. The whole possibility of applied science is implicit from the very start in the structure of theoretical modern science, which is a fundamentally technological way of knowing.

One sees this first in the transformation of the motive or end of philosophy, both natural and moral, enunciated by such thinkers as Bacon, Hobbes, and Descartes. The explicit purpose of theory is no longer disinterested contemplation but power. Speculative metaphysics falls into disrepute as the goal shifts from understanding and admiring the world to controlling it.[12] The fact that it was not until the nineteenth century that theoretical science actually began to yield practical fruits only makes more impressive these early declarations of technological intent.

In its method as well as its motive, the new science was inherently technological. This is visible, to begin with, in the very idea of "method," which now comes to the fore—the idea that the natural workings of the human mind are not to be trusted and must be replaced with a specialized, artificial technique for reasoning. Modern science does not, like earlier philosophy, grow out of and perfect common sense, raising it to "wisdom," but rather starts afresh, building a separate edifice of technical, "scientific" knowledge that sits alongside and in permanent tension with our natural way of knowing. Methodological science is a new thought technology.

The content of modern method, no less than its form, is also technological. Whether because of their new, practical motive, a change in metaphysics, or their new epistemology (this controversy need not be settled here), the very questions that such thinkers as Galileo, Bacon, and Descartes were asking, their whole conception of what would count as "understanding" nature, predisposed their thinking in a technological direction. Prior to all investigation, they tended to dismiss the older questions of "why" and "what" for the problem of "how." Henceforth, to understand nature would mean to know how it worked.

12. See Bacon, *The Great Instauration,* 4:24, and *The Wisdom of the Ancients,* 6:721, in *The Works of Francis Bacon,* ed. Spedding, Ellis, and Heath, 14 vols. (London: Longman, 1857–74); Hobbes *Leviathan,* ed. Michael Oakeshott (Oxford: Basil Blackwell, 1957), p. 435; Descartes, *Oeuvres et lettres,* ed. A. Bridoux (Paris: Pleiade, 1952), p. 168.

The precise method to be used in such research—which emerged not so much from any one of these thinkers as from their combined effect—had three central characteristics, each in its own way technological: it was analytical, mathematical, and experimental. By proceeding through analysis, by resolving every problem or phenomenon into (purportedly) more fundamental independent parts, scientific investigation rejected from the start, in its very method, the possibility that things must be understood as integral wholes (essences, formal causes, organisms) or understood in terms of some larger, overarching order or purpose (final causes). These excluded possibilities—that the world naturally tends toward the realization of certain structures or certain ends—would have placed a limit on both the legitimacy and possibility of the human manipulation of nature. Thus modern science's analytical or reductionist method, by denying the integrity of natural wholes, by requiring us to break down nature (at least in thought) in order to understand it, is inherently technological.

This point can be strengthened by considering an obvious objection. The ancient Epicureans, with their metaphysical materialism, also rejected formal and final causes, yet did not favor conquering nature but rather living in accordance with it. The difference derives not only from Epicurean moral doctrine but also from their belief in the irreducible heterogeneity of qualitatively distinct atoms, which puts strict limits on the malleability of nature. By contrast, the modern discovery of consciousness, or of subjectivity, or of the primary/secondary distinction placed man outside of nature, at least to the extent of being the repository of all secondary qualities; it thus drained external reality not only of noetic but of all qualitative or sensuous heterogeneity, leaving only an abstract and purely quantitative world of homogeneous matter. Consequently, whereas Lucretius explained the different qualities of things in terms of the inherent properties of their atoms, modern science sought to explain everything in terms of the different configurations or combinations of uniform matter. And since these configurations, once understood, might be recreated or transformed by man, this more radical reductionism or "analysis" opened a whole new vista on the manipulation of nature.

In breaking completely with the qualitative world and reducing everything to quantity and relation, modern science, in fact, went beyond mere analysis to a unique *mathematicization* of physics. Again, it is true that Pythagoras and, in some interpretations, Plato also saw mathematics as the key to the universe, but by this they meant that the cosmic whole has a permanent, static structure or form based on certain mathematical relationships. Modern science, by contrast, holds that the universe neither possesses nor tends toward fixed structures of any kind but is in constant, aimless flux. It is to the laws of this flux, and not to any static order, that modern science applies its mathematics. And whereas mathematical physics in the Pythagorean sense is clearly antitechnological and harmonistic, in the modern sense it is the very opposite. It means that, on the one hand, the world is vastly more changeable or malleable than it at first appears to be, but, on the other, that its constant change is not some incomprehensible Heraclitean flux but rather a precise, determinate, lawful sequence of cause and effect, susceptible to mathematical prediction and thus human control.

As analytical and mathematical, scientific method grounds the possibility of controlling nature; as "experimental," it proclaims the necessity for doing so. Modern science is built not on idle, speculative reason, nor even on empirical observation as such, but on the active manipulation of nature. The assumption is that nature, which is not ordered so as to serve the basic needs of human beings, is also not organized to serve their need to know. There is nothing privileged about nature's spontaneous self-presentation, nothing "more natural" about the forms it happens to assume in our particular environment. To extract its secrets, therefore, it is necessary to abandon contemplative detachment and passivity, to dissect, probe, and transform nature, breaking down through a kind of "analysis in action" in order to isolate its component parts for observation and measurement. In modern method, knowing is strictly inseparable from acting and manipulating.

As our knowledge has progressed, the reliance of theory on action has grown ever greater. It was not very obtrusive when science required no more than an inclined plane or simple telescope, but today, with the need for vast feats of invention and engineering, such as huge accelerators and interplanetary probes, one is struck by the utter dependence of theoretical knowledge on technological mastery. Modern science lives by the principle that one cannot understand the world without changing it.

In sum, the whole possibility of "applied science" and of the technological mastery of nature did not just happen to turn up as an innocent finding of theoretical science, an unanticipated discovery. It is not at all posterior to or separate from theoretical science but rather implicit from the very start in the latter's conception of "understand" and of the proper method—analytical, mathematical, and experimental—needed to pursue it. Modern science is intrinsically—intentionally and methodologically—technological.[13]

The modern convergence of the crafts and natural philosophy, then, does indeed involve the transformation of the latter no less than the former. And to the extent that modern science has become the reigning theoretical authority and the model for all genuine knowledge, it is possible to say that we are all prisoners of a technological consciousness. In its every dealing with the world, the modern mind tends to a kind of thinking that is technological in form, content, and motive.

Technology Bound and Unbound: The Idea of Tradition and the Idea of Progress

Modern technology is unique, we have seen, by virtue of its more intensive and extensive development, its spread to every aspect of life, its creation of a new dependence of humanity on technology and of nature on humanity, and its intimate, reciprocal relation to natural philosophy. Yet to understand the full meaning of technology, especially its constitutive role in the moral psychology of modern life, it is necessary to go beyond the more or less familiar themes just discussed to two further points, perhaps no less familiar but less adequately understood. These are the modern belief in progress and disbelief in happiness, or, in terms less misleading in their simplicity, the "futurity" and "negative orientation" of modern life.

In the first place, what we mean by "technology" is ultimately unintelligible without a proper conception of its link to that larger movement of which modern science is still only a part: faith in progress, the commitment to humanity's gradual self-liberation through

13. See Hans Jonas, *The Phenomenon of Life: Toward a Philosophical Biology* (New York: Dell, 1966), pp. 92–95, 200–207; and Alexandre Koyré, *From the Closed World to the Infinite Universe* (Baltimore: John Hopkins University Press, 1957). Consider also Heidegger, "Science and Reflection," in *Question concerning Technology.* For the opposite view, see Stephen Toulmin and June Goodfield, *The Architecture of Matter* (Chicago: University of Chicago Press, 1962), p. 39: "If, during the last few decades, scientists have at last been able to supplement, and even to revolutionize, the age-old craft traditions, this has been an incidental—and largely uncovenanted—fruit of scientific success."

its progressive appropriation of the natural and historical worlds. In the premodern era, after all, the various arts were in no sense conceived as a single historical force engaged in a long-term battle for the human conquest of nature. But this thought stands at the very core of the modern concept of technology.

Prior to the "age of progress," the arts were not seen as part of some great journey of mankind—a historical project or movement—but simply as individual crafts addressed to specific and limited needs, essentially set in their ways, even if susceptible to gradual improvement. Moreover, far from glorifying human freedom and power, the arts anxiously endeavored—for reasons to be considered momentarily—to obscure this as much as possible, attributing their origins to the gods, taking their models from nature, hedging themselves in with a hundred arbitrary rules and traditions, and embedding their practices in religious ritual and elaborate social structures (trade secrets, apprenticeships, guilds, and so forth). In short, in premodern or traditional society, the arts treaded quietly, outside the circle of people's deepest cares and loyalties, and indeed in bondage to them. This was the age of technology bound.

All of this changed with the rise of humanism and the idea of progress. All external and limiting conditions on the arts were gradually cast off, man became a conscious creator, "method" came to the fore as an art of discovery, "inventor" became a job, research and development were institutionalized, and the advance of human control became a deliberate, universal, open-ended project. It is this revolutionary change alone that gave birth to technology properly so-called, in which each specific art is experienced as part of the larger movement of all humanity toward freedom and mastery.

This great historical transformation is, of course, familiar enough, but its full significance for the uniqueness of modern technology cannot be seen if it is understood in the conventional way, as a triumph over the simple cowardice and superstition of the past by the *sapere aude* of the Enlightenment. The fact is that all premodern societies tend to be deeply suspicious of technological innovation and science. Although this tendency may stem in part from fear, prejudice, and ignorance, it is also a necessary concomitant of the whole structure and life orientation of such stories—of their rootedness in *tradition*. Only through the wholesale reordering of this comprehensive moral-political foundation could the idea of progress emerge and, with it, technology properly so-called. Thus this is the change we need to understand.

In all societies untouched by "modernization" or "Westernization," the foundation of the social world—the legitimacy of the government, the respectability of the social hierarchy, the authority of prevailing mores—is rooted in ancestral custom or tradition. But how does such a traditional society work? What is the source of the power of tradition over human minds and hearts? This power may be said to rest on three distinct forces, each of which will turn out to be in necessary tension with technological innovation.

The force of tradition derives, in the first place, from habit: we naturally develop a loyalty, fondness, and inclination for what we have become accustomed to over a long period of time. But the importance of habituation leads to an essential tension between custom or law and the arts, as Aristotle states in his classic discussion of whether laws should be as changeable as the arts: "The argument from the example of the arts is false. Change in an art is not like change in law; for law has no strength with respect to obedience apart from habit, and this is not created except over a period of time. Hence the easy alteration of existing laws in favor of new and different ones weakens the power of law itself" (*Politics* 1269a19–24). Since habituation requires changelessness, a healthy traditional society must be relatively static; therefore, it must necessarily be suspicious of the example and influence of the arts, with their reformist and innovative tendencies.

The second source of the power of custom is the communal consensus or public opinion in which it is embedded. The widespread judgment of society, especially of its oldest members, carries (not unreasonably) great authority, which the individual naturally tends to trust and respect in preference to his own limited powers of discernment. This rational deference to consensus is further supported and strengthened by the censorious pressure of public opinion, which rewards agreement and harshly punishes dissent. Custom, in other words, draws upon a certain natural inclination within groups toward intolerance, closed-mindedness, and conformity—an inclination that modern culture deplores as senseless, but that would seem to be essential (within measure) to the self-preservation of a traditional society, since its authority and world view are supported not by rational demonstration but only by the brute fact of actual agreement. Outside the communal consensus lies an abyss; and sensing this, people naturally hate what is strange. It is perhaps not impossible for such a society to maintain its self-certainty in the face of fundamental dissent or diversity, but it is always an ordeal. Consequently, reason and the arts, while certainly necessary to such societies, must always remain essentially suspect, owing to their tendencies to skepticism and innovation and their epistemological stance outside of consensual authority.

Finally, tradition exercises its power over people's minds because it is old or ancestral. Things very old, like those very large, naturally inspire awe because they dwarf our own limited extension in time and space. We also revere ancient customs because they represent the consensus of untold generations, as well as the product of a searching test of time.

Without further support, however, these considerations could not prevail against the debunking of tradition contained in the following reflection from Aristotle: "The first [human beings], whether they were earthborn or preserved from a cataclysm, are likely to have been similar to average or even simpleminded persons [today] . . . so it would be odd to abide by the opinions they hold" (*Politics* 1269a3-8). What truly vindicates the reverence for tradition, then, and constitutes its deepest ground is the belief—essentially religious—that our ancestors were vastly superior to us. There is a logic to this view. Just as parents are stronger and wiser than children (who also owe an infinite debt to their progenitors), so our ancestors were greater than we. What comes earlier must be superior to what comes later, since the former has created or begotten the latter. Indeed, thinking one's way back to the very beginning, one realizes that the divine beings of the first age, who were not the children of any earlier generation, generated the world itself and thus must have been absolutely the most powerful. In our own late and feeble age, this power has either diminished or withdrawn from the world, since changes of such magnitude no longer occur. But our earliest ancestors were close to this power, in a golden age when heroes strode the earth and humans conversed with gods. The ancient traditions, handed down to us from our ancestors are then to be cherished and revered. They come from a godlike source far greater and wiser than ourselves.

This foundational belief in the inherent superiority of the past to the present represents the third and greatest source of inescapable tension between the arts and traditional society. Technological innovation threatens not only fixity needed for habituation and for communal consensus but also the indispensable attitude of reverence for the ancestral. In other words, what it means, on the deepest level, for a society to be "traditional" is not merely that its subjects follow old customs but that, on some level, they understand themselves to have fallen from the originary fullness of the world; that they live in remembrance, repentance, and hunger for the past; and that they cleave to tradition as their sole tenuous link to the ever-receding, life-giving ancestral source. In this light, the arts must appear deeply suspect, for novelty is essentially betrayal. And our familiar idea of progress, could it arise, would seem altogether reckless, unholy, and insane.

In sum, in a society constituted on the basis of tradition, the opposition to reason and the arts does not derive from mere timidity or ignorance but reflects a permanent structural conflict between this open, dynamic, progressive aspect of human nature and the, as it were, ontological rootedness of communal life in the ancestral past. To employ a Nietzschean idiom, a fundamental tension exists here between thought and life. This is not to say, obviously, that the arts cannot exist in a traditional society, but rather that they do so precariously and only by virtue of the elaborate accommodations described above.[14] They are eternally gnawed upon by anxiety and guilt, as Prometheus would discover.

It follows, conversely, that in order for technology finally to be unbound, as in the modern era—in order for life to embrace reason, and humanity the ideal of progress—it is first necessary to supply an entirely new foundation for morality and politics. That foundation constitutes the indispensable, but long-forgotten ground of the modern turn to technology.

It was with this thought in mind that the Enlightenment consciously endeavored to subvert traditional society, to liberate humanity from all customary and religious authority, and to reconstitute the state on the basis of the free consent and enlightened self-interest of the resulting autonomous, rationalistic, and future-directed individual. In this new world, obedience would result not from the static

---

14. On these general themes, see Mircea Eliade, *Cosmos and History: The Myth of the Eternal Return*, trans. Willard Trask (New York: Harper & Row, 1959), and *Myth and Reality*, trans. Willard Trask (New York: Harper & Row, 1963); Leo Strauss, "Progress and Return," in *The Rebirth of Classical Rationalism: An Introduction to the Thought of Leo Strauss*, ed. Thomas Pangle (Chicago: University of Chicago Press, 1989); and Edward Shils, *Tradition* (Chicago: University of Chicago Press, 1981). See also Daniel Lerner, *The Passing of Traditional Society: Modernizing the Middle East* (Glencoe, Ill.: Free Press, 1958); and *Technology in Western Civilization*, ed. Melvin Kranzberg and Carroll W. Pursell, Jr. (New York: Oxford University Press, 1967), 1: pt. 1.

structure of habit, consensus, and reverence for the past but from the dynamics of the desire to get ahead—not from changelessness but precisely from the collective pursuit of change. Similarly, the legitimacy of the political hierarchy would no longer be based on some delicate communal consensus regarding ethics and religion but on the morally neutral principle of contract or consent. In such a society, which thrives on innovation, diversity, skepticism, and materialism—on everything most lethal to a traditional community—the technological impulse could at last be fully unbound. For the first time in history, it would no longer exist in fundamental tension with man's primary loyalties.

Indeed, technology now becomes not only permissible but, in the same moment, necessary. The social bond of enlightened self-interest—which, replacing that of tradition, permits the unleashing of technology—is in fact a bond at all, rather than a source of conflict, only insofar as the economy expands over the long run. And, in the absence of imperialism, the economy can expand steadily only if there are continual gains in productivity stemming from the advance of technology. In other words, in the modern liberal-democratic mass nation-state—the ultimate result of the Enlightenment's assault on traditional society—the diverse, deracinated, self-interested individuals, sharing no bond of "race, creed, or national origin," can live in harmony only through a certain hostility toward nature. We liberated men and women of the free society are united only by sharing the spoils of our common technological imperialism.

But there is still more. In the reconstitution of life engineered by the Enlightenment, technology becomes not merely released from its age-old bondage to tradition, and not merely an indispensable economic condition of the system that releases it: it is moved from the periphery to the very heart of what people live for. As the primary vehicle or expression of progress, it replaces tradition as the moral center of people's lives.

In this new society based on self-interest, after all, there is still idealism and morality. The idealistic citizen lives for the "getting ahead" not only of himself but also of larger communities. In the first place, this concern relates to his family—that is, the new, future-oriented, child-centered family. Such citizens live not for their ancestors but for their children, whom they consider in some sense their superiors. Beyond that, they seek to "be useful" and to "make a contribution"—but to what? Again, not to the ancestors, not even primarily to the present community, but to the future. They feel the need to have "a cause"; they want to "help build a better world." This is now what it means to be "idealistic." As always, it means devotion to a larger whole, but the whole is now not something extending back in time but forward—no longer a fatherland but a "movement."

The moral life necessarily takes the form of membership in a movement, which means it is rooted in the future and in the idea of progress. For the idealistic modern citizen, who he is, is where he is going—and not where he has come from. He lives, not in remembrance but in activist expectation, and understands himself, not as fallen from some golden age of primordial fullness but as part of the progressive march of civilization beyond the barbaric past (the state of nature) and toward a better future for all. To be sure, powerful vestiges of tradition remain, as do obligations to the present community, but these ties must always justify themselves as forces for progress. Modern patriotism loses legitimacy if one's nation is shown to be on the wrong side of history. In a manner unintelligible to the traditional citizen, the future of humanity now constitutes the highest good and the ultimate homeland of people's hopes and loyalties.

In other words, the replacement of the idea of tradition with that of progress means far more than that people now tend to believe that things will improve. It is a fundamental reversal of life orientation, in which the source or fullness of being is relocated from past to future. Human lives now derive their meaning and weight not from rootedness in the ancestral but from being the wave of the future. Modern man is a forward-looking, upwardly mobile, futural being whose profoundest suffering is no longer the guilty sense of fallenness or estrangement from the ancestral source but rather the fear of failure and the terror of being left out or behind.

In this transformed world, technology inevitably becomes the guiding and inspiriting force. We derive our sense of right from feeling that we are in touch with the future; and the most palpable sign of that contact is technology, which is, as it were, the most futural thing. It offers the surest proof that there has been progress and that there will be more; it is the clearest candidate for a world-historical force. Indeed, today, when one tries to form a picture of the distant future, one feels unsure about political and economic arrangements, but one always imagines it more technologically advanced. In our dynamic world, that is our one certainty: the future belongs to technology. That is why nothing is more deeply demoralizing and subversive in the present age than shoddy and outdated technology, as the experiences of Eastern Europe would seem to show. The good is the "advanced." In sum, in this new, future-directed world, it is now tradition that is fundamentally suspect and in tension with life; conversely, it is human art, reconceived as "technology"—a unitary, ever-advancing *movement* for human mastery—that stands at the center of life, the true spring of human vitality.

There is a common tendency to think of technology as a morally neutral and universal phenomenon whose modern development simply resulted from the advance of civilization—the gradual accumulation of discoveries, spurred on by curiosity and material need while resisted by human timidity and superstition. But the true character of technology, as indeed of modernity itself, cannot be grasped unless one sees that the technological turn was a culturally singular event made possible only by a unique moral, political, and, as it were, existential revolution that, overturning the entire traditional order, released the arts from their heretofore necessary bondage and, what is more, placed them at the heart not only of the political-economic system but of the whole purposive structure of modern life.

The Rejection of Happiness and the "Negative Orientation" in Life

Closely connected to the progressive, futural, movement-oriented character of contemporary existence is our typically modern skepticism concerning the possibility of rest, happiness, or a summum bonum. This skepticism, together with its result—the turn to a "negative orientation" in life, which seeks the avoidance of evil rather than the attainment of good—constitutes a second crucial but neglected element in the rise of modern technology.

One can see this most clearly as follows. The foregoing discussion of the transition from traditional to progressive society could be said to have omitted a middle term. Human life may be anchored in the ancestral past or eschatological future but also in that which is timelessly present: a rational, natural standard of the human good—of happiness or the summum bonum—as elaborated, for example, by Aristotle in the *Ethics*, and indeed by virtually all the competing sects of classical philosophy.

If one evaluates technology from this alternative standpoint—neither tradition nor progress but the natural human good—one comes across the remarkable fact that virtually all of the ancient schools, Platonists as well as Aristotelians, Stoics no less than Epicureans, agreed in opposing the unregulated development of the arts, the endless pursuit of mastery, and unlimited acquisitiveness. Indeed, two millennia before Rousseau, most of them entertained a "primitivist" skepticism about the real value of civilization and its progress.[15]

To some extent these attitudes simply reflected an acknowledgment of the requirements of traditional society, for the "Greek enlightenment," in contrast to that of the eighteenth century, never envisaged the wholesale replacement of such a society with one based on reason. This aside, their reasoned conception of the good was also antitechnological. It was argued that knowledge—the perfection of the mind—must surely have a worthier and more satisfying purpose than finding comforts for the body. Indeed, the intrinsic joys of contemplation soon produce disgust with all such vulgar utilitarianism. Even the greatest ancient technologist, Archimedes, held to this view (as reported by Plutarch): "He would not deign to leave behind him any commentary or writing on such subjects; but, repudiating as sordid and ignoble the whole trade of engineering, and every sort of art that lends itself to mere use and profit he placed his whole affection and ambition in those purer speculations where there can be no reference to the vulgar needs of life" ("Life of Marcellus," p. 378). Even the Epicureans such as Lucretius, who rejected the life of contemplation and identified the good with pleasure, shared this contempt for the arts and acquisition. What, after all, can technology really do for us? The true needs of nature are very few and easily satisfied. Pleasure, like all things, has its natural limits; beyond that, all is vanity and illusion. If one chases after power and possessions in the hope of satisfying all of one's desires, one will only succeed in increasing toil, strife, and the dependence on fortune and one's fellow man, while not gaining one's contentment; such desires are infinite and always run ahead of our power to satisfy them. It is better, then, to put desire and power in harmony, not by extending the latter but by limiting the former—by moderation, *the* classical virtue.[16]

Of course, the same conclusion is reached if, like the Stoics and

15. On the subject of ancient primitivism, see Arthur Lovejoy and George Boas, *Primitivism and Related Ideas in Antiquity* (Baltimore: Johns Hopkins University Press, 1935). See also Arthur Melzer, *The Natural Goodness of Man: On the System of Rousseau's Thought* (Chicago: University of Chicago Press: 1990), pp. 55–57.

16. See James H. Nichols, Jr., *Epicurean Political Philosophy: The De rerum natura of Lucretius* (Ithaca: Cornell University Press, 1972), pp. 101–78. See also Nichols, Chapter 1 in this book.

others, one identifies the good with moral virtue. In fact, so long as one has *any* conception of the good or final end, regardless of its content, one will reject as utterly senseless the endless pursuit of means.

It is crucial to note, therefore, that modern thought begins precisely with the rejection of the possibility of happiness or a summum bonum. As Hobbes proclaims in a famous passage, "The felicity of this life, consisteth not in the repose of a mind satisfied. For there is no such *finis ultimus,* utmost aim, nor *summum bonum,* greatest good, as is spoken of in the books of the old moral philosophers. . . . Felicity is a continual progress of the desire, from one object to another; the attaining of the former, being still but the way to the latter" (*Leviathan,* p. 80). There is nothing the *possession* of which brings satisfaction. Hence, there is "felicity," the pleasure attending the ceaseless motion from one temporary goal to another, but no happiness—no contentment, completion, and repose. Life is endless wanting and striving.

But, then, striving for *what?* If nothing brings satisfaction or happiness, then for what purpose and with what hope do we make something the goal of our striving? According to Hobbes, there can be no positive goal of human striving but only a negative one: the avoidance of evil, especially the greatest and most comprehensive evil, death. His radical "non-eudaemonism" entails a fundamental reversal of life orientation: the pursuit of good is replaced with the overcoming of evil, the summum bonum of happiness with the summum malum of death.

But the avoidance of death or the attainment of security is an unfinishable task, and therefore the proper and inevitable character of human desire is "a perpetual and restless desire of power after power, that ceaseth only in death" (ibid.). In this way, the pursuit of means and power was liberated from all concrete ends. Through this momentous revolution, technology, already unbound from the fetters of tradition, was cut loose from those of the human good as well—set free to become, for the first time, an independent goal and indeed the central object of philosophy, society, and progress.

What Hobbes wrought has not yet been undone. The negative orientation toward life, founded on the dismissal of happiness, continues to dominate modern culture, which does not hold up for emulation any image of the completed or perfected life. We believe in goods, to be sure, in various pleasures and comforts, but not in a complete good or ultimate purpose that might structure a way of life. In the absence of such a comprehensive positive goal, the lives of individuals and of nations tend to be organized negatively by the avoidance of death and the derivative evils of war, hunger, poverty, disease, oppression, intolerance, and superstition. These evils, and not some vague talk about happiness, are what we really believe in. We define progress in terms of the distance traveled from them (or from the state of nature, barbarism, the Middle Ages, the old country). Lacking positive goals, knowing what we are against but not what we are for, we tend to cherish opportunity over accomplishment, rights over goods, liberation over consummation, youth over maturity, hope over fulfillment, and, above all, means over ends. Consequently, we embrace technology with an abandon that would not be possible if there were clarity regarding the ultimate end. What might be called the "forgetfulness of happiness" thus lies at the core of the technological turn.

SOME PROBLEMS OF TECHNOLOGY

Having examined the character of modern technology and the technological character of modernity—with special emphasis on the futurity and negative orientation of modern life—we are in a position to assess some of the more prominent formulations of technology's "problem." Our foremost concern is to know whether this problem deserves the position it has recently assumed as the central or most fundamental philosophical issue.

The Irony of Technology

All serious conceptions of the problem presuppose some form of the thesis that technology—which seems to be a mere means and thus a derivative phenomenon, subordinate to human choice—has become, in fact, an independent force that rules its putative master. The seductive irony of this thought has encouraged much exaggeration, but it is difficult to deny that there is much truth to it.

First, there is the fact that as soon as one chooses a means, one begins to obey it—to do the myriad things necessary for its acquisition, maintenance, and employment. Think of all the ways we are ruled, for example, by our use of oil as a means. To put it in Marxist terminology, the "means of production" determine the "relations of production"; or, as Rousseau insisted, nothing is more enslaving than the pursuit of power, for to control, one must obey.

Second, every choice of means usually ends up subjecting us to important unintended consequences that we did not and would not choose, such as pollution or global warming. One might, of course, reply to this point, as to the first, that we still remain free in that we can choose or reject the whole package of a given means with all of its requirements and consequences. But in practice the full contents of the package are seldom learned until much later, and by then the means may be so built in to our world that we hardly remain free to reject it. When the automobile was invented, we had the power but not the knowledge to rule it; later we acquired the knowledge but lost the power. And that is how it *must* be with technological progress, which brings a constant stream of new, unknown forces.[17]

Third, we lose our freedom to control a given means not only because it integrates itself into the world but also into ourselves. Objects we first choose as luxuries soon become necessities. Similarly, things like television do not remain external to us, where we are free to judge them, but gradually alter our very tastes and perceptions. In general, as the world becomes more technological, we become more technological—in our sensibilities, our modes of thinking, perhaps in our very posture toward Being. Our judgment is transformed by the very thing we are trying to judge. Thus, toward so pervasive and transforming a force as technology, we are unable to maintain a truly external and free relationship.

Finally, choices regarding technology are not made by single agents in isolation but in a system with other, competing players. Frequently we must either develop and adopt new technologies or lose all—militarily or economically—to others who do. It is often the case that no one would choose a given technique for its own sake—dangerous and expensive weaponry or an efficient but environmentally harmful manufacturing process—and yet the system forces all to do so. In a competitive system, inventions rule their makers.

In sum, technology appears to be more than a mere collection of means, subordinate to human choice, but is an independent force with its own logic or destiny to which humanity is compelled to

---

17. See Introduction by Kass in this volume.

submit. This claim forms the starting point for all conceptions of technology as the most determinative or fundamental phenomenon.

Although there is clearly an element of truth to this claim, evident from the sorts of general arguments outlined above, we are perhaps too inclined to think that the issue can be settled on the level of these kinds of arguments. In fact, it would be an enormously complex philosophical-sociological-historical undertaking to try to establish just how independent and controlling a force technology is in practice. Without entering into this question further here, I would simply observe that no one has yet succeeded in answering it convincingly or even in showing how one should go about answering it.[18]

A related question, which remains on our more general level, is what the inner logic or destiny of this more or less independent force might be and how it renders technology the central problem. Returning to the rough typology employed at the beginning, we can distinguish three prevalent answers: a popular or common-sense view and those of the radical left and right.[19]

The Popular Problem of Technology

In the public mind today, technology is seen as a relatively independent force, having as its primary logic a simple tendency toward expansion: over time, technology multiplies the kinds and magnitude of human interference with nature. And this is seen to be a problem, owing primarily to the corresponding growth of unintended consequences—ozone depletion, deforestation, contamination of food, air, and water, and so forth.

But practical dangers of this kind, however urgent they may be, do not lead to the raising of serious theoretical questions about technology as such. There may be some general hand-wringing of a philosophical nature, but when people finally set themselves to the task of finding a solution, their primary hopes soon turn back in the direction of new technology (for example, solar energy). Indeed, for this reason the problem of unintended consequences, far from undermining the technological project, actually strengthens it, giving people

18. For a good discussion, see Langdon Winner, *Autonomous Technology: Technics-out-of-Control as a Theme in Political Thought* (Cambridge: MIT Press, 1977).

19. The distinction between "left" and "right," never very clear, becomes less so with each passing year. It is employed here because it remains more useful than any equally simple distinction that could be put in its place.

the sense that it is no longer really an option to stop, that now more than ever we need the enlightened technology of tomorrow in order to save us from the dangers of today's.

Even in their motivation, people who react to problems of this kind remain essentially technological. Now that everyone sees technology as an independent force able to do us harm—indeed, a force more alien and harmful than nature itself, which now seems our friend—it inevitably occurs to them to bring it, too, under human direction and control. No one proposes that we should submit to such historical forces with resignation and acceptance. We must master them. The mainstream movements of environmentalism, antinuclearism, and others are simply the modern technological attitude now extended to the force of technology itself.

The Problem of Technology on the Left

On the popular or common-sense level, the problem of technology remains a technological problem. It becomes a political and philosophical one only on the basis of more determinate views of the destiny of technology, especially as it relates to prevailing political forms. The left argues that it is quite impossible to meet the above dangers of unintended consequences within the confines of a capitalist, pluralist, representative liberal democracy.

First, capitalism is so utterly dependent on technology, the sine qua non of continued growth, and so immersed in its ethos that it is incapable of recognizing its dangers, let alone dealing with them effectively. Second, in a pluralist society given over to competing material interests, there is no institution that looks to the good of the whole, nor any interest group or lobby for the future; in a representative democracy, unborn generations have no vote.[20] In such a system, where could the political force come from that would protect the environment or preserve resources for the future? Thus only a socialist society—planned, humanistic, and wary of growth—possesses the ability to tame technology.

But, third, if we do not succeed in thus taming technology and freeing our minds from its mode of thought—instrumental rationality—we will be led by its irresistible logic to the ever greater "rational-

20. Jonas, *Philosophical Essays*, pp. 18–19; Robert Paul Wolfe, "Beyond Tolerance," in *A Critique of Pure Tolerance* (Boston: Beacon Press, 1970), pp. 49–52.

ization" and organization of society, the transformation of human beings into tools and cogs, the loss of all individuality and spontaneity, the rise of technocratic elitism, the strengthening of the state through new technologies of military force, surveillance, information processing, mass media, and propaganda, and, in sum, the creation of a dehumanized technological despotism.[21]

While for many years it seemed almost axiomatic that capitalism was in thrall to technology and that the latter was a force hostile to liberty, these views have recently come to sound strangely old-fashioned and out of touch. The third argument, in particular, seems difficult to sustain in light of the undeniable world trend toward democracy and, what is more, the important contributions to this trend made precisely by the new information and communication technologies. More generally, today the logic of events seems to be pushing away from social regimentation or mechanization and toward individual initiative as the source of cooperation and energy, away from central planning and toward local responsibility and experimentation, thus also away from socialism—which suddenly seems the more narrowly technological of the two—and toward free-market spontaneity. The great temptation today is actually to invert the old left-wing view and argue that the inner logic of technology leads to democracy and capitalism. But from this striking shift of perspective a more reasonable conclusion would be that there is probably no such logic at all, and that, over the long term, the social and political effect of technology will continue to change in unknowable ways.[22]

The first two arguments also seem to have been refuted by events. It is undeniable that the contemporary environmental movement first

21. See especially Erich Fromm, *The Revolution of Hope: Toward a Humanized Technology* (New York: Harper & Row, 1968); Lewis Mumford, *The Myth of the Machine,* 2 vols. (New York: Harcourt, Brace, Jovanovich, 1967–70); Marcuse, *One-Dimensional Man;* Theodore Roszak, *The Making of a Counter Culture: Reflections on the Technocratic Society and Its Youthful Opposition* (Garden City, N.Y.: Doubleday, 1969); Horkheimer, *Eclipse of Reason.* See also Wolin, Chapter 7 in this book.

22. On this practical level, one must also question the too-easy equation of technology and science. Is there not truth in the older Enlightenment view that, whatever effect technology may have at the moment, the underlying scientific enterprise requires and fosters certain moral habits favorable to individual liberty and democracy, such as skepticism, dissent, tolerance, creativity, open debate, objectivity, and love of truth? See the classic statement by Jacob Bronowski, *Science and Human Values* (New York: Harper & Row, 1965). See also Paul Goldstene, *The Collapse of Liberal Empire: Science and Revolution in the Twentieth Century* (New Haven: Yale University Press, 1977), pp. 91–127; and Yaron Ezrahi, *The Descent of Icarus: Science and the Transformation of Contemporary Democracy* (Cambridge: Harvard University Press, 1990).

arose in the United States and other Western democracies and every year becomes more widespread and powerful. There is a nontrivial sense in which today almost everyone in the United States is an environmentalist, albeit with varying degrees of fervor and influence. So how is this possible in the land of interest-group politics and the capitalist ethos? Where is the flaw in the left-wing critique? What resources could exist within liberal capitalist culture for taming technology? Some answers emerge on the basis of the particular characterization of modernity and technology developed above.

It is a great, if common, mistake to assume, as the left-wing critique does, that in a pluralist liberal democracy all interest groups must be concerned with immediate, selfish goods. On the contrary, in modern societies, where life is grounded in the idea of progress, a concern for the future forms some part of everyone's thinking, and future-directed, ideological interest groups tend to proliferate. Certainly there is no structural reason why an environmental movement could not arise in a liberal society. To see why it became, in fact, the ideological growth stock of the 1980s and beyond, one might consider some further factors.

As we have seen, when technology comes to be viewed as an independent and dangerous force, the will to control it inevitably emerges from the technological attitude itself. Technology is, as it were, a self-regulating movement. And the fear of technology, once aroused, actually fits perfectly into the liberal capitalist world view, strange as that may seem. Skeptical regarding the good and happiness, liberal societies tend to measure progress negatively, in terms of the triumph over evil—over the barbarism and backwardness of the state of nature. But the negative orientation of the liberal mind has one inevitable shortcoming: due to the very progress of society, these natural evils gradually recede, losing their power to ground and orient our lives. New evils must be found to take their place. That is, I believe, the psychological and existential meaning of liberal environmentalism. Ecological disaster has taken the place of the state of nature. The middle classes of the advanced industrialized nations, who can no longer feel the old bourgeois sense of pride in the triumph of civilization over barbarism (and who have seen the demise of fascism and communism), have redefined progressivism in terms of saving the planet.

Yet from the perspective of left-wing common sense, it still seems that this should not be possible. Liberal capitalism embraces a world of competition and strife, not of harmony with nature. Wholly committed to growth, technology, and the unbridled exploitation of nature, it contains no countervailing principle that might reign in humanity. It simply lacks the resources to check the human quest for power.[23]

The error here is to assume that liberalism's hostility toward nature entails an unlimited trust in humanity. Liberalism's deepest root is precisely a fear of humanity: we are not by nature sociable or good for each other; our original condition is a state of war; government has been invented to reign in humanity; and this invention itself must be carefully restrained through popular representation, separation of powers, constitutionally grounded rights, and so forth. In the economic realm as well, there must be no monopoly, whether public or private, but rather free competition and mutual checking. The core of liberalism, in short, is a distrust of all unchecked human power. (This contrasts with the socialist tradition, with its greater trust in the goodness of man and of human power, so long as the latter is centralized and public.) Thus it is not unintelligible or even surprising that our traditional ethos of limited government should have expanded in recent decades to include the need for limited technology.[24] As shocking as it may be to long-accepted ideas, the liberal capitalist psyche turns out to be uniquely fertile soil for the growth of a certain kind of environmentalism—that based on the negative idealism of triumphing over disaster, on the deep suspicion of human power, and on the will to control technology resulting dialectically from the technological impulse itself.

The foregoing arguments are not meant to prove, however, that technology necessarily leads to liberal democratic capitalism or that the latter must inevitably succeed in taming the former. They only seek to call into question the long-prevalent theory that these outcomes are essentially impossible. They do further suggest, however, that the fortuities of moral and political life, not the automatic unfolding of technology's supposed inner logic, will determine whether

---

23. See Roszak, *Making of a Counter Culture*, and Merchant, *The Death of Nature*.

24. This argument, of course, makes no sense in the left-wing interpretation of liberalism, according to which the principle of limited government was never anything but a secret means for *removing* limits from economic power. It is true that liberalism prefers to regulate economic activity through market rather than governmental mechanisms and that it regards economic power as less dangerous than political, but that does not mean that liberals are insincere in their distrust of human power as such.

freedom or despotism will ultimately triumph, or unintended consequences be controlled.[25]

The Problem of Technology on the Right

There is a further set of problems, however, that seems more
strictly tied to the phenomenon and fate of technology as such. Cultural conservatives, rooted in both the religious and the romantic-
existentialist right, have long opposed technology as inevitably leading to aesthetic, moral, and spiritual degradation (themes also picked
up, of course, by the left and especially the new left).

Even if environmental disaster is avoided and every trace of oppression eliminated, the argument goes, the unhindered success of technology necessarily leads to the expansion of the material world, the
deepening spread of consumerism, the endless production of new
goods and needs; and all of this distracts and seduces human beings
from their higher vocation, threatening the permanent atrophy of
their nobler faculties. Universal affluence and ease will cause people
to become spoiled and decadent, robbing humanity of the indispensable lessons of adversity, and fostering a world of lethargic and un-
disciplined mass men. At the same time, all the beauties of external
nature and all the living artifacts of traditional culture will be razed to
make room for new factories, condos, and shopping malls—a sterile
landscape to accompany an empty heart.[26]

Still more deadly to the soul than these material effects of technology, however, is the underlying thought or attitude. Amid even the
greatest natural splendors, the mind feels little if it sees only the
inanimate world of modern physics. What true belief can there be in
romantic love or political greatness on the basis of our reductionist
psychology, what notion of human freedom and dignity given our
determinist natural and social science? No lofty human possibility
can flourish in the disenchanted, leveled, homogeneous, quantitative
world of modern scientific rationalism. And yet, the argument continues, this life-destroying rationalism is not imposed on us by the na-

---

25. See Galston and Weinberger, Chapters 10 and 11, respectively, in this book.

26. See, for example, Ortega y Gasset, *Revolt of the Masses*; Henri Bergson, *The Two Sources
of Morality and Religion*, trans. R. Ashley Audra and Cloudesley Brereton (New York: Holt,
1935), pp. 298, 302–9. See also Borgmann, *Technology and the Character of Contemporary Life*;
Alexis de Tocqueville, *Democracy in America*, trans. J. P. Mayer (New York: Doubleday, 1966),
pt. 2, bk. 2, chaps. 10, 11, 16, 17; bk. 3, chaps. 17, 19; bk. 4, chap. 6.

ture of reality itself but is rather something we impose on ourselves through our will to technological mastery. We will this nihilistic world without quality, form, or hierarchy, for such the world must be if we are to place it entirely at our disposal. Thus, in the end, it is the technological attitude that has stifled our souls by expelling from the world every possible experience of beauty, wonder, and reverence.

These two general views may be said to represent, respectively, the materialist and idealist accounts of the emergence of the soulless "mass man." Since the rise of European fascism, political theorists have not pushed such arguments with the old passion (except on the left), but there is still a good deal of common sense and popular opinion on this side. Certainly among third-world and other external observers, it is almost axiomatic that wealth and materialism have made the West spoiled, soft, shallow, and decadent. And we our-selves seem fairly obsessed by the thought that our brave new world is creating a rebarbarized population of yuppies, couch potatoes, and valley girls.

Yet the question that needs to be raised here is not whether these evils are a danger but only whether they follow inevitably from tech-nological progress, so that the "problem of technology" may be de-clared the crucial, underlying political and philosophical issue.

As to the materialistic theory of mass man, most observers would concede that—like the left-wing critique—it has not, as yet, been proved by the facts. Despite several generations of historic levels of affluence, the Western democracies continue to show some strength of character and have thus far been able to meet the challenges of all those presuming upon their irreversible decadence.[27] One possible explanation for this, which meets the materialist argument on its own terms, is that the rapid advance of technology brings in its wake a great deal of social and economic disruption, inevitably producing almost as much adversity as it removes. Thus, if modern man has not yet sunk into the bovine happiness of the herd—disappointing long-standing expectations—it may be because, with the concomitant

27. In *The Present Age,* Kierkegaard writes that his generation has lapsed "into complete indolence. Its condition is that of a man who has only fallen asleep towards morning: first of all come great dreams, then a feeling of laziness, and finally a witty or clever excuse for remaining in bed. . . . In the present age a rebellion is, of all things, the most unthinkable. Such an expression of strength would seem ridiculous to the calculating intelligence of our times" (trans. Alexander Dru [New York: Harper & Row, 1962], p. 35). This was written two years before 1848. See also Solzhenitsyn's predictions for the outcome of the Cold War in his "Harvard speech," *A World Split Apart.*

erosion of neighborhoods and families, the rise of crime and divorce, and the constant threat of economic obsolescence, he does not yet feel terribly spoiled. One might also argue, on a somewhat brighter note, that capitalism, with its much decried tendency to engender competition, has for this reason proved remarkably resourceful in spurring energetic activity and accomplishment (not to say workaholism) in the very midst of prosperity. At least thus far.

Looking to the longer term, one could, of course, point to the fact that, in the past, leisure classes have existed that were not corrupted, but rather liberated and ennobled, by great affluence. To be sure, this required the existence of a strenuous, uniform, socially enforced, and usually aristocratic moral code, which is unlikely to return in our liberal, pluralistic, egalitarian, materialist culture. But the laws of the human spirit are still so unknown that no one can say with confidence that some new basis—perhaps religious or mystical—cannot emerge for the ennoblement of leisure.[28] In every age, moreover, there have been at least isolated individuals who, owing perhaps to a certain innate strength of mind or character, were able to rise above their circumstances and answer the call of their highest possibilities. And it is not unlikely that technological progress will someday give us the power—through drugs or genetic engineering—to produce that innate strength and thus liberate humanity from the influence of the otherwise stunting cultural conditions that such material progress also brings.[29]

As to the idealist theory of mass man, which puts the blame on scientific rationalism stemming from technology, one could respond with a point similar to that made above regarding liberal environmentalism: this romantic-existentialist critique of technology is, in fact, a dialectical expression of the technological attitude itself. Consider its origins.

Enlightenment rationalism and modern science, which arose with the hope of controlling nature for the sake of man's physical well-being, had the unintended consequence of undermining his psychic or emotional well-being. He felt homeless and estranged, his spirit unable to breathe, in the depleted metaphysical environment created

28. See Bergson, who argues precisely that technology will lead to mysticism (*Two Sources of Morality and Religion*, pp. 300–303, 308–17).

29. "The mind depends so much on the temperament and on the disposition of the organs of the body that if it is possible to find some means which generally renders men wiser and more skillful than they have been hitherto, I believe that it is in medicine that one must search" (Descartes, *Oeuvres*, pp. 168–69).

by modern physics. But this predicament suggests a number of possible responses. One could have accepted this bleak world with resignation and the renunciation of human mastery—either on the model of Pascal, for example, or of Lucretius, who found a new, austere happiness by adjusting his soul to the harsh truths of ancient atomism.

Instead, however, the problems of modern science were addressed in the modern spirit of indignation and rebellion, as if humanity had a right to a cosmos that supported its noblest passions. Our whole understanding of the "problem of technology" was itself formulated from the standpoint of man's quest to master all the conditions of his existence. And consequently its solution was sought through a further recourse to technology. In Romantic thought, the artistic side of human nature was elevated above the rational: man's essence was relocated in creativity. It is within man's power and right, it was argued, to engineer a new interpretation of nature and reason that would answer to the needs of the heart just as the old paradigm had to those of the body. Thus when the Romantics objected to the scientific-utilitarian exploitation of nature, it was ultimately only because this spoiled it for their own use—as a vehicle for aesthetic experience, a neutral "other" on which to project and unfold the inner self. The scientist sees nature as matter, the Romantic as subject matter. Both equally seek control, the one over the physical world of objects, the other over the subjective world of perceptions and feelings. Similarly, Nietzsche and the existentialists assert man's capacity to create himself, to legislate values, and to interpret the world in the service of human needs and aspirations.[30] In short, the problem of science and of the mass man is solved by extending technological control into the realm of what we now call "belief systems."[31]

If this solution is correct, however, then the problem of mass man derives not from technology as such but only from its incomplete application. Technology comes to light once again as a self-correcting

30. This, of course, is where Heidegger turns away from and against Nietzsche and existentialism. Without trying to judge whether Heidegger himself is successful in his attempt to free himself from technology, I would suggest that most of those today who attack technology in his name, in fact, remain closer to Nietzsche. Also see Vattimo (Chap. 9 here), who argues that Heidegger, remaining within the old "machine paradigm" of technology, failed to see the ontologically liberating possibilities of other technological forms, especially in the area of media and communications.

31. See Carl Schmitt's *Political Romanticism*, trans. Guy Oakes (Cambridge: MIT Press, 1986). For a discussion of related ideas see Jeffrey Herf, *Reactionary Modernism: Technology, Culture, and Politics in Weimar and the Third Reich* (Cambridge: Cambridge University Press, 1984), pp. 1–17, 109–29. See also Cantor, Chapter 5 in this book.

phenomenon—hence, not as the central political and philosophical problem.

On the other hand, one may question the adequacy of this solution. If we follow it and free our minds from the seemingly stifling constraints of scientific rationalism, perhaps a more elevated and authentic culture will eventually flower, but perhaps mass culture will triumph all the more fully. The tribunal of reason may hold us back, but it also holds us up. And the mass man, who chafes under every form of authority and discipline, may celebrate the demise of "truth" by sinking still further into bovine self-satisfaction. Liberated from bondage to reason, he is now free to believe whatever he wants—whatever makes him feel good about himself—because, as the ultimate expression of modern technology, he has placed his own beliefs at his disposal.

I am not arguing, of course, that either of these two scenarios is inevitable, but only that both are possible. Once more, it seems that the inner logic and future consequences of technology are far more indeterminate than its critics tend to suppose, and that independent moral and intellectual forces will continue to play a crucial role in shaping human destiny.

CONCLUSION

From this overview of the characteristics and problems of modern technology one cannot formulate firm conclusions, but it will not be inappropriate to suggest some general lessons and directions for thought.

We have seen that modern technology is indeed highly problematic, but it poses many different problems, some of them mutually exclusive, and it seems impossible to say which will loom largest in the future. Paradoxically, the more one surveys the large and fairly booming field of technology criticism, the more one is driven to conclude that the meaning and destiny of technology are indeterminate. And if this conclusion is correct, then there simply *is* no single issue that can be identified—and placed ceremoniously at the center of our thought—as the essential "problem of technology."

This is not to deny that technology is probably the most characteristic and imposing feature of our age, but for that very reason we are probably too inclined to attribute to it a specific, coherent, and all-determining significance. We should be more on our guard. If it is the true task of philosophy to liberate itself from the prejudices of the age, then perhaps our specific duty is to free ourselves from technology—not only from the deification but also the demonization of it (the true *idée fixe* of our time).[32] In addition to being theoretically questionable, moreover, this demonization—the excessive emphasis on the "problem of technology"—brings a specific problem of its own. When technology is viewed as the central and defining philosophical issue, elementary moral and political distinctions tend to lose their clarity and force. Liberal democracy, communism, and fascism, for example, appear fundamentally equivalent, since they are equally technological. Furthermore, to the extent that we view ourselves as helpless pawns of an overarching and immovable force, we may renounce the moral and political responsibility that, in fact, is crucial for the good exercise of what power over technology we do possess. Thus an exaggerated philosophical concern with "the problem of technology" hinders the effective management of the various real problems of technology.

At or near the heart of the modern technological turn, as I have suggested, are two things: the idea of progress and the radical subordination of ends to means resulting from the dismissal of the idea of a summum bonum. We might begin to distance ourselves from technology, then, by questioning these two things—first, by attempting to overcome our forgetfulness of happiness and the human good; and second, by adopting, as part of this very attempt, a more humble attitude toward the past, especially the pretechnological past, both Eastern and Western. In such earlier thinkers as, say, Confucius and Aristotle, one finds a comprehensive theory of happiness formulated in serene detachment not only from the turbulent hopes of modern technology but, just as importantly, from the angry rebellion against it. With their help, we might perhaps begin to free our minds from Western modernity and, at the same time, from modern self-loathing, of which "the problem of technology" is the greatest manifestation.

The same conclusion also emerges as follows: It is often said that

32. It will, of course, be objected that it is impossible to escape the ideas of one's time, but this familiar thesis has never been proved conclusively and probably cannot be. Thus, given our necessary uncertainty on this score, it seems best to act as if we were capable of freedom and see what develops, since to assume the opposite is to make the thesis self-fulfilling through a preemptive surrender to prevailing beliefs.

early modern philosophy was polemical, in that it arose not from a direct encounter with life but from a reaction against existing ideas—against "the kingdom of darkness from vain philosophy [i.e., Aristotle] and fabulous traditions [Christianity]" (in Hobbes's famous phrase). This reaction, moreover, concerned not only the falseness of the prevailing ideology but especially its practical consequences, its use as a tool of exploitation. Most of later modern thought has remained polemical—and indeed technological—in this sense: it grows not from a naive pursuit of truth but from our famous posture of alienation—a rebellion against the *consequences* of prevailing ideas and an activist determination to control and overcome them. Thus we speak of the "counter-Enlightenment," the "Romantic reaction," and so forth. And certainly most of contemporary philosophy, especially all variants of postmodernism, self-consciously derive from a political, moral, or aesthetic reaction against the consequences of instrumental rationalism—against "the problem of technology." But the more we rail against technology, the more we remain in its grip.

What seems to be needed is rather a nonpolemical form of thinking that takes its start not from a settled opinion regarding the danger of prevailing beliefs but from a more naive, honest, and direct encounter with life—an encounter that, following the model of an earlier, pretechnological mode of thought, draws upon the elemental and permanent human concern for happiness. Doubtless, there are problems with technology, but we will never adequately comprehend or respond to them if we do not begin by looking away from them to questions that are currently less urgent and imposing but are perhaps more honest and fundamental.

Notes on Contributors

PAUL A. CANTOR is Professor of English at the University of Virginia. He is the author of *Shakespeare's Rome: Republic and Empire* (1976), *Creature and Creator: Myth-making and English Romanticism* (1984), and *Shakespeare: "Hamlet"* (1989).

G. A. COHEN is Chichele Professor of Social and Political Theory at Oxford University and a Fellow at All Souls College. His first book, *Karl Marx's Theory of History: A Defense* (1978), won the Isaac Deutscher Prize. His other works include *History, Labor, and Freedom: Themes from Marx* (1988).

WILLIAM A. GALSTON is currently Deputy Assistant to the President for Domestic Policy. He is on leave from the University of Maryland, College Park, where he is Professor in the School of Public Affairs and Senior Research Fellow in the Institute for Philosophy and Public Policy. He is the author of *Kant and the Problem of History* (1975), *Justice and the Human Good* (1980), and, most recently, *Liberal Purposes: Goods, Virtues, and Diversity in the Liberal State* (1991).

STANLEY L. JAKI is Distinguished University Professor at Seton Hall University. He has been Gifford Lecturer at the University of Edinburgh (1974–75 and 1975–76) and Farmington Lecturer at Oxford University (1988). His books include *The Relevance of Physics* (1966), *The Road to Science and the Ways of God* (1978), *Uneasy Genius: The Life and Work of Pierre Duhem* (1984), and *God and the Cosmologists* (1989).

LEON R. KASS is Addie Clark Professor at the University of Chicago, where he teaches in the College and the Committee on Social Thought.

323

A trained physician and biochemist, Kass is the author of *Toward a More Natural Science: Biology and Human Affairs* (1985).

WILSON CAREY MCWILLIAMS is Professor of Political Science at Rutgers University. His works include *The Idea of Fraternity in America* (1973), *American Government* (1987), and numerous essays on the theory and practice of American politics.

ARTHUR M. MELZER is Associate Professor of Political Science at Michigan State University. He is the author of *The Natural Goodness of Man: On the System of Rousseau's Thought* (1990). Melzer is a Director of the Symposium on Science, Reason, and Modern Democracy.

JAMES H. NICHOLS, JR., is Professor of Political Science at Claremont McKenna College. He has published a book on classical materialism, *Epicurean Political Philosophy: The "De rerum natura" of Lucretius* (1976) and translations of Alexandre Kojève's *Introduction to the Reading of Hegel* (1980) and Plato's *Laches* (in *The Roots of Political Philosophy: Ten Forgotten Socratic Dialogues*, ed. Thomas Pangle [1987]).

STANLEY ROSEN is the Evan Pugh Professor of Philosophy at Pennsylvania State University. He has published books on Plato, Hegel, Heidegger, nihilism, analytical philosophy, and hermeneutics. Many of his essays have been collected in *The Quarrel between Philosophy and Poetry: Studies in Ancient Thought* (1988) and *The Ancients and the Moderns: Rethinking Modernity* (1989).

REINER SCHÜRMANN is Professor of Philosophy at the Graduate Faculty of the New School for Social Research. He is the author of *Heidegger on Being and Acting: From Principles to Anarchy* (1987) and *The Public Realm: Essays on Discursive Types in Political Philosophy* (1989). He has also published *Meister Eckhart: Mystic and Philosopher* (1978), which contains translations and commentary.

GIANNI VATTIMO is Professor of Philosophy at the University of Turin. *The End of Modernity: Nihilism and Hermeneutics in Postmodern Culture* (1988) and *The Transparent Society* (1992) are the first of his books to appear in English. His other books include *Le avventure della differenza* (1980), *Al di là del soggetto: Nietzsche, Heidegger, e l'ermeneutica* (1981), and *Introduzione a Nietzsche* (1985).

JERRY WEINBERGER is Professor of Political Science at Michigan State University. He is the author of *Science, Faith, and Politics: Francis Bacon and the Utopian Roots of the Modern Age* (1985) and editor of *Francis Bacon: "New*

Atlantis" and "The Great Instauration" (1989). He is a Director of the Symposium on Science, Reason, and Modern Democracy.

SHELDON S. WOLIN has taught at the University of California at Berkeley, the University of California at Santa Cruz, and Princeton University. His works include *Politics and Vision: Continuity and Innovation in Western Political Thought* (1960) and *The Presence of the Past: Essays on the State and the Constitution* (1989).

M. RICHARD ZINMAN is Professor of Political Theory in James Madison College at Michigan State University. He is Executive Director of the Symposium on Science, Reason, and Modern Democracy.

Index

327